AME OMY

Brock/Springer Series in Contemporary Bioscience

Nitrogen-fixing Bacteria in Nonleguminous Crop Plants

Brock/Springer Series in Contemporary Bioscience

Series Editor: *Thomas D. Brock*
University of Wisconsin-Madison

Tom Fenchel
ECOLOGY OF PROTOZOA: The Biology of Free-living Phagotrophic Protists

Johanna Döbereiner
Fábio O. Pedrosa
NITROGEN-FIXING BACTERIA IN NONLEGUMINOUS CROP PLANTS

Forthcoming in this series:

Tsutomu Hattori
THE VIABLE COUNT: Statistical and Quantitative Aspects

Johanna Döbereiner
Fábio O. Pedrosa

Nitrogen-fixing Bacteria in Nonleguminous Crop Plants

With 12 Figures and 24 Tables

Science Tech Publishers
Madison, Wisconsin

Springer-Verlag
Berlin Heidelberg New York
London Paris Tokyo

Dr. Johanna Döbereiner
Director, UAPNPBS
EMBRAPA
Rio de Janeiro, Brazil

Prof. Fábio O. Pedrosa
Department of Biochemistry
Federal University of Parana
Curitiba, Brazil

Library of Congress Cataloging-in-Publication Data
Döbereiner, Johanna.
Nitrogen-fixing bacteria in nonleguminous crop plants.
(Brock/Springer series in contemporary bioscience)
Bibliography: p.
Includes index.
1. Micro-organisms, Nitrogen-fixing. 2. Azospirillum.
3. Plant biotechnology. I. Pedrosa, Fabio O., 1947–
II. Title. III. Series.
QR113.D63 1987 589.9′504133 87-12901
ISBN 0-910239-11-8

Science Tech, Inc., 701 Ridge Street
Madison, Wisconsin 53705, U.S.A.

Sole distribution rights outside of the USA, Canada, and Mexico granted to
Springer-Verlag Berlin Heidelberg New York London Paris Tokyo

ISBN 0-910239-11-8 Science Tech, Inc. Madison, WI
ISBN 3-540-17969-0 Springer-Verlag Berlin Heidelberg New York

Production and editorial supervision: Science Tech Publishers
Design: Katherine M. Brock
Cover photograph: Electron micrograph of *Azospirillum*, magnification 15,000×. Reproduced from Krieg and Döbereiner, 1984, by permission of Williams and Wilkins.

Printed in the United States of America
10 9 8 7 6 5 4 3 2 1

Preface

After I moved to Brazil from Europe in the early 1950s I began my research by searching for diazotrophic bacteria which may occur in association with plant roots. I was motivated to study this topic by observing the continuously high-yielding sugarcane crops and other grasses of Brazil which required no fertilization. This observation drastically contradicted my previous education in Europe and convinced me that tropical agriculture must have many unique features, especially because the heavy rains and rapid decomposition of organic matter in the tropics must cause great losses of nutrients, especially nitrogen.

At that time several review articles (such as Allison 1947) had just appeared, which concluded that neither *Azotobacter* nor other free-living, diazotrophic bacteria can proliferate in the rhizosphere of plants, so that the contribution of these organisms to soil fertility is insignificant. Therefore, we received a skeptical response in the early 1960s when we reported selective multiplication of *Beijerinckia* on roots of sugarcane and the existence of a new *Azotobacter* species in evergreen Paspalum grass (Döbereiner 1961; Döbereiner 1966).

Many years of work were required to prove that economically significant levels of nitrogen can be fixed under field conditions. Were it not for the moral support of R. H. Burris, F. J. Bergersen, W. D. P. Stewart, P. S. Nutman, and others, we probably would soon have given up this research. Fortunately, however, increasing numbers of students came to our laboratory at the Brazilian National Research Program on Soil Biology: Among them was Fábio de Oliveira Pedrosa, the coauthor of this book, who is now on the faculty of the University of Parana, Curitiba, Brazil. In addition, from 1970 to 1971 while I was on a sabbatical leave at Rothamsted Experimental Station in England, Peter Dart and John Day joined the "Grass-N_2-Fixation Club," of which I was already a member. There we tried the acetylene reduction method using grass roots flown in from Brazil, and I am sure the three of us will never forget the day when a batch of bottles containing roots under acetylene, which had been forgotten for three days on the bench, suddenly showed tremendous ethylene peaks.

But there were many C_2H_2-reducing grass root samples which contained neither *Azotobacter* nor *Beijerinckia*. The first observations of the frequent occurrence of vibrio-like organisms with very characteristic cork-screw movements, which seemed to be associated with C_2H_2-reducing root pieces, were made in our laboratory in 1973 by J. J. Peña (unpublished data). But for more than a year the importance of this organism remained unrecognized, because cultures of it reduced C_2H_2 only when mixed with contaminants, just as Beijerinck observed in his early experiments (1925), that *Spirillum lipoferum* could fix N_2 only in mixed enrichment cultures. Also, Becking (1963) observed $^{15}N_2$ incorporation only when yeast extract was added to the medium. The introduction of N-free semisolid media was finally the key which allowed the identification of our isolates as true microaerobic diazotrophs, and as the *Spirillum lipoferum* described by Beijerinck in 1925.

In the paper we delivered at the 1st International Symposium on Nitrogen Fixation in Pullman, Washington, in 1974 (Döbereiner and Day 1976) five important characteristics of tropical nitrogen fixation were suggested, which seemed difficult to believe at the time: (*a*) *Azospirillum* is a very common, widespread soil and root inhabitant; (*b*) the reason for the success of the use of semisolid media in the isolation of microaerobic diazotrophs is its O_2 gradient; (*c*) root tissues are infected by diazotrophs; (*d*) plant-diazotroph interactions exist; and (*e*) N_2 fixation in association with grasses can reach economically significant levels. Today most of these hypotheses have been confirmed and expanded further as will be discussed in this monograph.

Due to the importance of expanding N_2 fixation to cereals, many research approaches have been tried, especially as the "green revolution" now appears to be unrealistic, due to the energy crisis in tropical countries which are mostly oil importers. The two main currents of research, which both started about 10 years ago, are: the conventional current, which tries to identify and improve already existing associations of diazotrophs with cereals and grasses; and the futuristic current, which hopes to create new diazotrophic symbioses from the many possibilities which are now offered by the rapid progress in genetic engineering. We will have to wait and see which current will be more successful. Although, in the long run, new manmade plants seem to offer the more promising solution, there also seems to be a place for short- and medium-term solutions which can use the many new and recently discovered diazotrophs and their associations with plants. These organisms are the result of natural evolution which has allowed the survival of both the microbes and the plants.

During the last ten years there have been several ups and downs in research on the conventional current and many serious scientists still do not believe in the potential benefits for agriculture of such associations. A few research teams, however, including ours, have continued research firmly along these lines and we are certain that, especially in tropical countries, agriculture will benefit substantially from the results of these research programs.

The many recent, encouraging results described in this book demonstrate the progress and extent of work in this field and, above all, are intended to show how little we yet know of these associations and how important it is to better understand them in order to make them commercially useful.

Finally, this book would be incomplete without the chapters on *Azospirillum* physiology and genetics, which were written very expertly by Fabio Pedrosa. He once was my student, but today is my teacher on these subjects. This work also owes much of its substance to our discussions with various colleagues, especially with Robert M. Boddey, Vera Lucia D. Baldani, Jose Ivo Baldani, Segundo S. Urquiaga C., and with many of our students.

Fabio Pedrosa and I would also like to thank the following: The United States National Academy of Sciences BOSTID program and the Brazilian National Research Council (CNPq) for financial support and Janette Teixeira for typing the manuscript in a very professional manner.

Johanna Döbereiner
Rio de Janeiro, Brazil

Contents

Brock/Springer Series in Contemporary Bioscience

Nitrogen-fixing Bacteria in Nonleguminous Crop Plants

1

Introduction

Biological dinitrogen fixation ($N_2 \longrightarrow NH_3$) is the single most important alternative for the increasingly high costs of N fertilizer input into cropping systems without substantial loss in yield. Crop rotation with legumes and diversification of agricultural systems offer many possibilities for attracting increasing attention, especially from those who feel responsibility for the environment. Still, there is a growing pressure for additional alternatives, and the extension of N_2 fixation to cereals and forage grasses remains a major challenge for agricultural research.

Because higher plants and even unicellular eucaryotes are not known to be able to use molecular N_2 in nature, attempts at transferring the N_2-fixing ability to higher plants have involved procaryotic microorganisms (Giles and Whitehead 1975; Merrick and Dixon 1984). An exciting possibility is the involvement of associations of certain *Agrobacterium* strains with dicotyledenous plants, resulting in the formation of tumors due to genetic exchanges between plant and bacteria. Such a system opens up possibilities of transferring genes responsible for the complex ensemble of N_2 fixation, and O_2 protection, into higher plant cells. Other possibilities arise from the similarity of RNA sequences of *Enterobacteriaceae* with those of chloroplasts (Merrick and Dixon 1984), which could be engineered to carry out simultaneous photosynthesis and N_2 fixation, as in the cyanobacteria.

A more conventional approach seems to be the identification and possible manipulation of already existing associations of plants with N_2-fixing bacteria. Associations of specific bacteria with plant roots are not new. The best known examples of such associations are pathogenic in nature and are usually readily recognized by damage to plant tissues.

The vast literature on plant pathology is useful in general terms in that it provides a basis for concepts in root bacteria associations (see e.g., Panopoulos *et al.* 1984). For example, it is well established that plant cultivars or species vary in their susceptibility to certain microorganisms. Many other kinds of plant–microbe interactions exist, with varying degrees of specificity, in the soil environment. Even in plant pathology, the various mechanisms responsible for specificity are not well understood, but present-day knowledge of plant and microbial genetics makes their elucidation possible.

Until 1950, the only well-defined free-living N_2-fixing bacteria were aerobic *Azotobacter* spp. and the anaerobic *Clostridium pasteurianum* (Waksman 1952; Winogradsky 1949). The results of many attempts to use these organisms as inoculants for cereals were not reproducible (Allison 1947), and these failures were attributed to the inability of *Azotobacter* to establish itself in the rhizosphere in reasonable numbers (Rovira 1963), especially under temperate climatic conditions. In the tropics, where most soils are acid, *Azotobacter* spp. are only found sporadically because of their high pH requirements. The only experiments in which azotobacters were found to occur abundantly in soil and in the rhizosphere were carried out in Egypt (Abd-el-Malek 1971).

In predominantly acid tropical soils *Beijerinckia* spp. were found to predominate (Becking 1961a; Derx 1953; Döbereiner and Campelo 1971). Initial evidence of associative N_2 fixation in Gramineae involved experiments with sugarcane in Brazil. Selective enrichment of *Beijerinckia* spp. in the sugarcane rhizosphere was observed (Döbereiner 1961), and this appeared to be the first indication of a change in the microbial equilibrium in the rhizosphere in favor of N_2-fixing bacteria. A few years later, a specific association of a new *Azotobacter* sp. (*A. paspali*) with one tetraploid cultivar of *Paspalum notatum* was demonstrated (Döbereiner 1966, 1970). Modification of the rhizosphere population in favor of a N_2-fixing *Bacillus* sp. by the substitution of specific pairs of chromosomes in wheat was later reported by Neal and Larson (1976) and Larson and Neal (1978).

Following the discovery of the C_2H_2 reduction method for determinations of nitrogenase activity, associative N_2 fixation was demonstrated in many grasses and cereals (Boddey and Döbereiner 1982; Neyra and Döbereiner 1977; Van Berkum and Bohlool 1980). Several additional N_2-fixing bacteria were isolated from rhizosphere soil, from root surfaces, and from within roots (e.g., *Pseudomonas* spp. and *Enterobacteriaceae*), where they were often found in very high numbers (Barraquio and Watanabe 1981; Pedersen *et al.* 1978; Rinaudo 1977). Proper identification of these organisms, however, is often difficult and their capacity to fix N_2 seems variable. Most of them are unable to grow with N_2 as the sole N source, making their isolation more difficult. This, however, is not a proof of their lack of importance. In fact, ***Rhizobium*** spp., the legume symbionts, which are by far the most important N_2-fixing bacteria, show similar characteristics. The transference of genetic information among various bacteria in the rhizosphere, where these organisms find proper conditions for rapid multiplication, may in fact be involved in the frequent occurrence of the capacity to fix N_2 of certain strains of species that would normally not be expected to do so. A possible explanation for this may be found in the observation that in some organisms in this group and in rhizobia, the *nif* genes are localized on plasmids.

Most reports, however, dealing with N_2 fixation associated with grasses and cereals consider azospirilla to be the major responsible diazotrophs. Because this new group of diazotrophs shows higher and more constant N_2 fixation and is easier to identify, most bacteriological work has been concentrating on azospirilla. In this monograph the relevant work on root-associating diazotrophs, their association with plants, and their agronomic importance has been reviewed.

2

The Genus *Azospirillum*

Beijerinck (1925) observed a very characteristic "spirillum" in enrichment cultures in N-free liquid medium and suggested that N_2 fixation in mixed cultures were attributed to this organism, which he named *Spirillum lipoferum*. Schröder (1932) isolated this organism from a large number of soil samples from throughout Germany and Austria, but could not confirm N_2 fixation in pure cultures. Dinitrogen fixation by an "aerobic vibrio or spirillum," which was suggested as being possibly identical to *S. lipoferum*, could be confirmed by the incorporation of $^{15}N_2$ gas in liquid medium containing yeast extract (Becking 1963). More than a decade later, however, it became clear that this organism could fix N_2 only in cultures with a N starter dose or where the O_2 availability was limited, either by contamination with O_2 scavenging microorganisms or by restricted O_2 diffusion in semisolid media (Döbereiner and Day 1976). In such cases, the respiration rate of the bacteria balances that of O_2 diffusion so that the dissolved O_2 level remains near zero and O_2 does not accumulate within the medium.

Using semisolid N-free media, with malate as the carbon source, it became very easy to enrich and isolate *Spirillum*-like N_2-fixing bacteria from the rhizosphere soil and from grass and cereal roots. These bacteria were then identified as *Spirillum lipoferum* according to the description of Beijerinck (1925). More detailed studies on a large number of isolates from various parts of the world led then to the description of a new genus named *Azospirillum*. This originally consisted of two species, *A. lipoferum* (syn. *S. lipoferum*) and *A. brasilense* (Tarrand *et al.* 1978), and recently two more species, *A. amazonense* (Magalhães *et al.* 1983)

(a) (b) (c)

Figure 2.1 Characteristic growth sequence of *Azospirillum* in semisolid media in a test tube. (a) Initial outgrowth from the inoculation point. (b) Growth into a balloon-like form. (c) Final formation of a pellicle.

and *A. halopraeferans* (Reinhold *et al.* 1987), have been proposed and confirmed by DNA/DNA and DNA/rRNA homology experiments.

2.1 Characterization of the genus

In semisolid N-free media (Döbereiner 1980) the commencement of growth manifests itself as a veillike pellicle or balloon, often 10 mm or more below the surface (Figure 2.1). Formation of this pellicle is due to an aerotactic response of the motile bacteria toward the levels of O_2 concentrations that permit N_2 fixation (Barak *et al.* 1982; Okon 1985; Okon *et al.* 1980), that is, where the dissolved O_2 concentration is just enough for optimal respiration rates without inhibiting N_2 fixation (Day and Döbereiner 1976). In semisolid media with combined nitrogen, similar veils or balloons are formed, although optimal growth occurs under air in liquid media, confirming the chemotactic attraction to a specific low O_2 tension. When growth continues and more O_2 is needed the pellicle moves toward the surface where a dense, wavy, paperlike pellicle forms. This growth pattern in semisolid media is one of the most characteristic features of the growth of azospirilla and permits its identification in enrichment cultures.

Azospirillum spp., as defined in *Bergey's Manual of Systematic Bacteriology* (Krieg and Döbereiner 1984), have curved plump rods 0.8 to 1.0 μm in diameter and 2 to 4 μm in length, often with pointed ends.

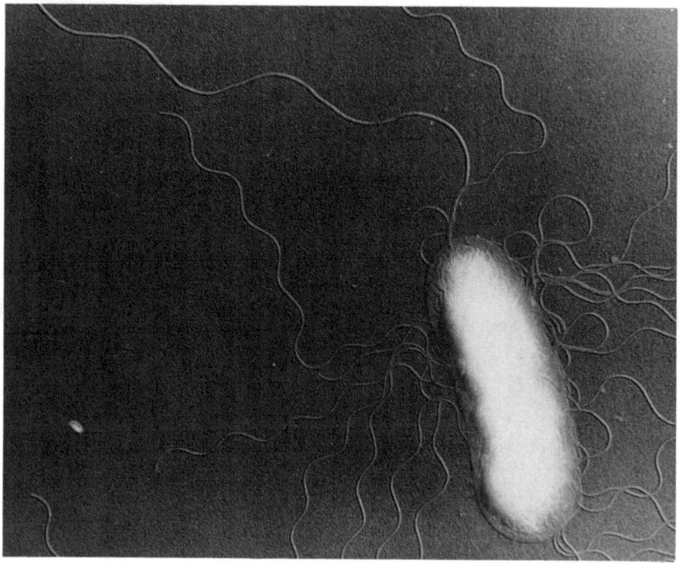

Figure 2.2 Electron micrograph of an *Azospirillum* cell grown on soft nutrient agar, showing both the single, polar flagellum and the finer, lateral flagella (15,000×). (Reproduced from Krieg and Döbereiner 1984 by permission from Williams and Wilkins.)

Intracellular granules of poly-β-hydroxybutyrate (PHB) are formed which, in N_2-fixing cells, can represent as much as 50% of the cell dry weight, while in cells cultured with NH_4^+ salts they constitute only 1% of the cell (Okon *et al.* 1976b). Enlarged polymorphic forms may occur in old alkaline cultures or under conditions of high O_2 tensions. The cells are normally Gram-negative, but Gram variability has been observed with *A. brasilense* (Tarrand *et al.* 1978). The organism is very motile in semisolid or liquid media and has a characteristic corkscrew-boring or vibratory motion; it is propelled by means of a single polar flagellum (Figure 2.2). In liquid cultures, peculiar spiral structures have been found associated with the flagellum (Nurmiaho-Lassila *et al.* 1981). Numerous lateral flagella of shorter wavelength (Figure 2.2) may be observed in soft nutrient agar where swarming is observed (Hall and Krieg 1983) and where there are antigenic differences between lateral and polar flagella (Hall and Krieg 1984). Filamentous forms have been observed in liquid media containing beef extract or high concentrations (7 g/liter) of yeast extract (Becking 1985). Nonmotile, enlarged pleomorphic cystlike forms (called C-forms) may occur (Figure 2.3) in *A. lipoferum* and *A. brasilense*, especially in older cultures (Eskew *et al.* 1977; Tarrand *et al.* 1978) on nutrient-limited agar plates containing

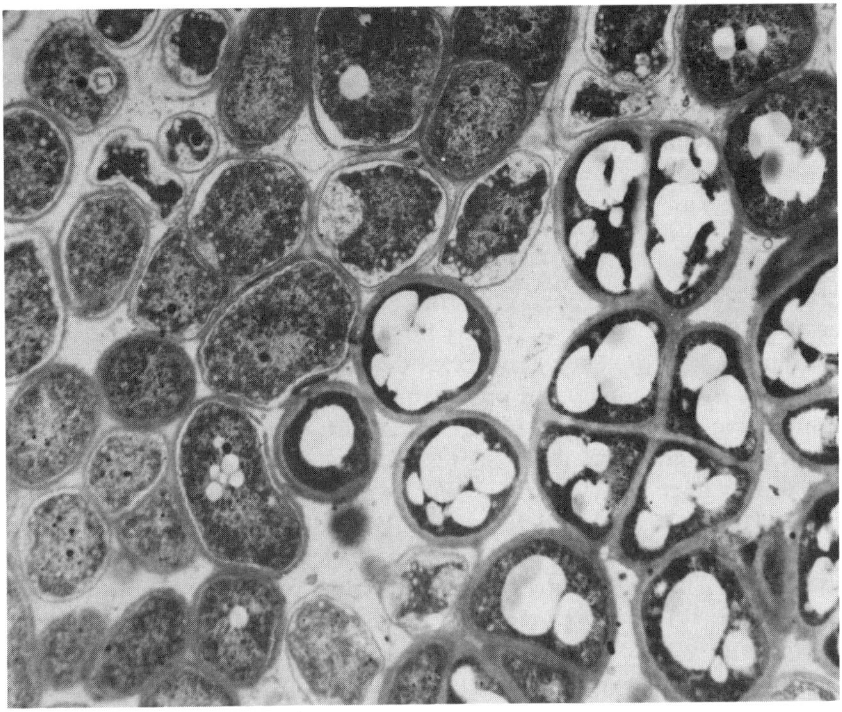

Figure 2.3 Ultrastructure of vibrioid and encapsulated forms (C-forms) of *Azospirillum brasilense* grown in association with a sugarcane callus. (Reproduced from Berg *et al.* 1979 by permission from Springer-Verlag.)

fructose (Berg *et al.* 1980; Sadasivan and Neyra 1985), in association with grass roots (Umali-Garcia *et al.* 1980), or in plant tissue cultures (Berg *et al.* 1979) where they have been suggested to serve for protection of nitrogenase against O_2 damage. The ultrastructure of the C-forms has little similarity with that of *Azotobacter* cysts, although Lamm and Neyra (1981) and Sadasivan and Neyra (1985) showed that C-forms have greater resistance to desiccation than vegetative forms and therefore should be considered true cysts.

In liquid media such cysts can become embedded within a fibrillar matrix, forming flocs (Becking 1985). On lake water agar *A. lipoferum* forms conglomerates surrounded by layers of acidic polysaccharides (Falk *et al.* 1986; Skerman *et al.* 1983).

Azospirilla are typical aerobes, when supplied with a combined N source, and possess mainly a respiratory type of metabolism with O_2 or NO_3^- as the terminal electron acceptor, but weak fermentative ability has been observed with *A. lipoferum* grown on glucose or fructose

(Tarrand *et al.* 1978). Optimum growth temperatures are between 35 and 37°C. The organism is oxidase positive and chemoorganotrophic, and some *A. lipoferum* strains can be facultative hydrogen autotrophs (Sampaio *et al.* 1981). Azospirilla grow well on organic acids such as L-malate, succinate, lactate, pyruvate, and transaconitate. Fructose, galactose, and some other sugars are also used as carbon substrates; many require biotin. The mole percent G + C of the DNA is 67 to 71. The type species is *Azospirillum lipoferum*.

A comparison of the *Azospirillum* spp. is shown in Table 2.1. Three distinct species have been confirmed by DNA homology studies (Falk *et al.* 1985; Tarrand *et al.* 1978) but the extent of DNA/rRNA hybridization (Falk *et al.* 1986) indicates that their relationship is sufficient to justify their inclusion in one genus (Table 2.2). The description of four species, according to Krieg and Döbereiner (1984), Magalhães *et al.* (1983), and Reinhold *et al.* (1985a), is as follows.

The first species is *Azospirillum lipoferum* (Beijerinck 1925), Tarrand *et al.* 1978, syn. *Spirillum lipoferum* Beijerinck 1925. The morphology is as depicted in Figure 2.4*a* (page 11) and as described for the genus. The characteristic pleomorphism in *A. lipoferum* (Figures 2.4*b* and 2.4*c*) occurs in media containing organic acids as the carbon source, and which turn alkaline following growth. Such enlarged cells remain viable and seem to have unchanged cell walls (De-Polli and Roth 1981). In young cultures or in those grown in media with sugars as carbon sources and which do not become alkaline, pleomorphic forms do not develop. This pleomorphism is the most readily recognizable feature for the routine differentiation of *A. lipoferum* from the very similar *A. brasilense*. The use of glucose and the requirement of biotin serve as additional differentiation tests.

All strains are uniformly Gram-negative when cultured in nutrient broth. On nutrient agar, a small portion of the cells is resistant to Gram decolorization. Colonies on potato agar containing malate are visible after 48 hours and are small, white, round, and smooth. After 1 week they progressively become dry and wrinkled with varying shades of pink and are characteristically indented into the agar. This characteristic colony type occurs only when the medium turns alkaline. When the malate is replaced by glucose or fructose, large, slimy, white colonies are formed. Aerobic growth in liquid media containing a source of mineral nitrogen is rapid, and extensive clumping of cells is often observed. Strains of *A. lipoferum*, which are able to grow lithoautotrophically with H_2, contain an uptake hydrogenase as well as ribulose 1,5-diphosphate carboxylase (RubPcase). They can be classified as aerobic, hydrogen-oxidizing bacteria. Physiological and nutritional characteristics are described in Table 2.1. The mole percent G + C of the DNA is 69 to 70. The type strain is ATCC 29707 (strain 59b of Tarrand *et al.* 1978), and other

Table 2.1 Comparison of the four *Azospirillum* spp.[a]

	A. brasilense	*A. lipoferum*	*A. amazonense*	*A. halopraeferans*
Cell width (μm)	1.0–1.2	1.0–1.5	0.8–1.0	0.7–1.0
Pleomorphic cells in alkaline media	−	+	−	+
Colony type on potato agar	Pink, raised, curled	Pink, raised, curled	White, flat, raised margin	No growth
Dissimilation of[b]				
NO_3^- to NO_2^-	+	+	±	+
NO_2^- to N_2O	±	±	−	+
NO_3^- dependent anaerobic growth	+	+	−	nd
Biotin requirement	−	+	−	+
Use of glucose[c]	−	+	+	−
Use of sucrose, maltose, trehalose, lactose	−	−	+	−
Use of α-keto glutarate	−	+	−	+
pH range for good growth	6.0–7.8	5.7–6.8	5.7–6.5	6.8–8.0
DNA base composition (mol % G + C)	69–70	69–70	66–67	68.1

[a]From Krieg and Döbereiner 1984, Magalhães *et al.* 1983, Reinhold *et al.* 1987, and Martinez-Drets *et al.* 1985.

[b]Symbols: +, all strains tested; ±, several strains positive but majority negative; −, all strains negative; nd, not determined.

[c]D-fructose, D-galactose, L-arabinose, lactate, malate, succinate, pyruvate, and fumarate are used by all four species. Malonate, D-erythrose, D-arabinose, L-rhamnose, cellobiose, and CM-cellulose are not used by *A. brasilense* or *A. lipoferum*.

Table 2.2 DNA and RNA homologies among *Azospirillum* spp. and *Herbaspirillum seropedicae*[a]

	Homology with reference strains (%)[b]			
	A. lipoferum	*A. brasilense*	*A. amazonense*	*H. seropedicae*
A. lipoferum DNA/DNA	69	14	6	15
RNA/RNA	99	91	65	13
A. brasilense DNA/DNA	16	100	7	23
RNA/RNA	96	100	68	15
A. amazonense DNA/DNA	4	3	92	6
RNA/RNA	76	66	100	21
H. seropedicae DNA/DNA	7	2	1	69
RNA/RNA	20	20	18	100

[a]From Baldani *et al*. 1986c, and Falk *et al*. 1985, 1986.

[b]Numbers show means of several strains.

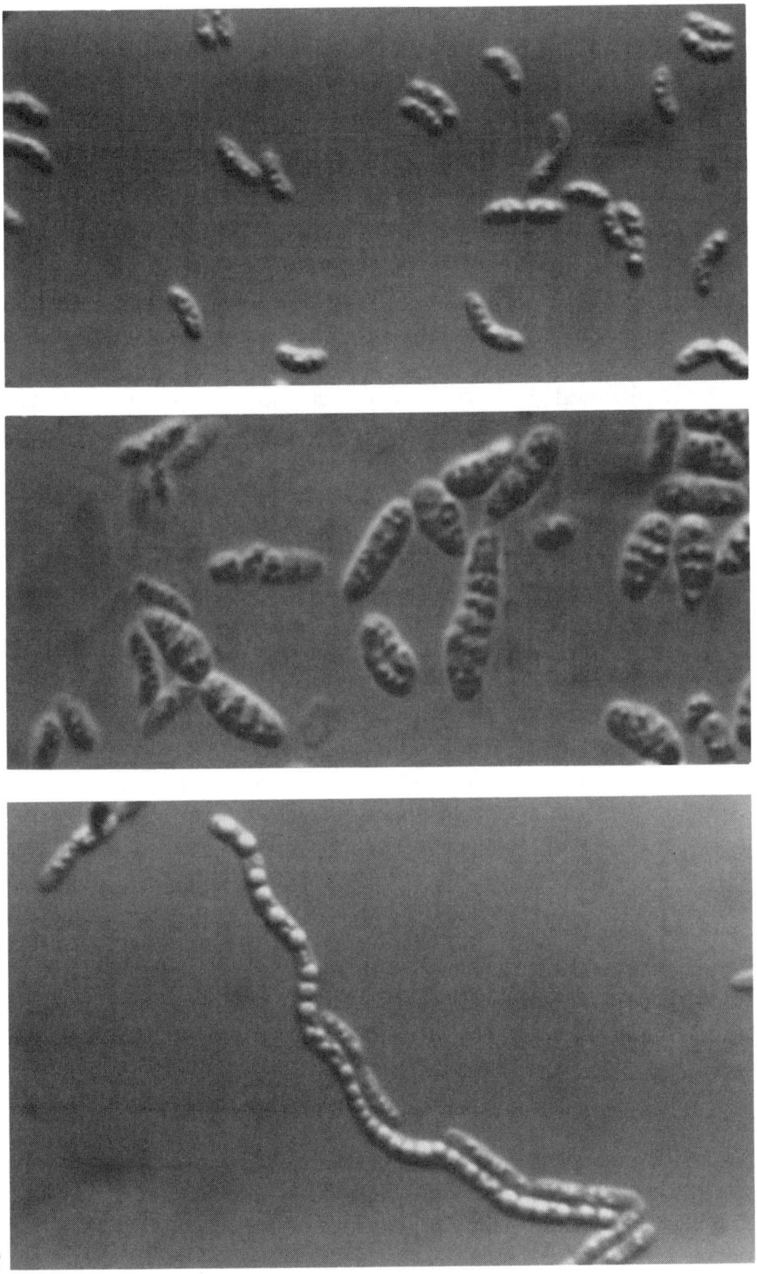

(a)

(b)

(c)

Figure 2.4 (a) Young, vibrioid cells of either *A. lipoferum* or *A. brasilense*, which are indistinguishable at this stage, containing PHB granules (2,100×). (b) and (c) Pleomorphic older (10-day) cells of *A. lipoferum* (2,100×). (Courtesy H. De-Polli.)

reference strains are ATCC 29708 (strain SpRG 20a), ATCC 29709 (strain SpBr17), and ATCC 29731 (strain SpRG6xx).

The second species is ***Azospirillum brasilense*** Tarrand, Krieg and Döbereiner (1978) (pertaining to Brazil), which has a morphology as depicted in Figure 2.4*a*. It is indistinguishable from young cultures of *A. lipoferum*, but the cells remain vibrioid and highly motile even in older alkaline cultures. All strains stain uniformly Gram-negative when cultured in nutrient broth, but most of them exhibit a small proportion of cells having resistance to Gram decolorization when cultured on nutrient agar. This Gram variability may be related to encapsulated cysts (or C-forms). Colonies on potato agar are indistinguishable from those of *A. lipoferum*, but bright red mutants may occur. Growth in liquid media with mineral N is relatively fast, homogeneous, and without clumping. The physiological and nutritional characteristics of the species were presented in Table 2.1. The mole percent G + C of the DNA is 70 to 71. The type strain is ATCC 29145 (strain Sp7 of Tarrand *et al.* 1978). Other reference strains are ATCC 29710 (strain Cd forming red colonies) and ATCC 29711 (Sp35).

The third species is ***Azospirillum amazonense***, Magalhães *et al.* (1983) (pertaining to the Amazon region). Cells are typical azospirilla with spinning movements, but are slightly smaller than the other species (0.9 by 3.5 μm) with prominent PHB granules in N_2-grown cultures (Figure 2.5*a*), while NH_4-grown cells are slimmer and have only small granules (Figure 2.5*b*). The lateral flagella observed in the two other species on nutrient agar (Tarrand *et al.* 1978) do not occur and there is also no swarming. Growth in N-free malate medium (pH >6.5) is scant, but is abundant in slightly acid sucrose or glucose media where the pH does not increase. Optimal pH for growth and N_2 fixation is in the range of 5.8 to 6.8 and at higher pH values viability is decreased. Addition of chelated Fe compounds did not relieve its sensitivity to alkaline reactions. Colonies on potato agar are characteristically flat and white, with elevated margins (Figure 2.6). The optimum growth temperature is 36°C with growth limits at 25 and 40°C. Nitrogenase activity is more sensitive to O_2 than in the two other species. Nitrate is assimilated but not dissimilated or denitrified, and there is no anaerobic growth in the presence of NO_3^-. Sucrose and glucose are favored carbon substrates and there is no fermentative ability. The mole percent of DNA is 67 to 68. The type strain is ATCC 35119 (strain Am 14 or YI) and ATCC 35120 (strain Am 18); ATCC 35121 (strain Am 30) are additional reference strains.

A comparison with the other *Azospirillum* spp. is presented in Table 2.1. Characteristics for routine identification are the scant pellicle formation in NFb medium containing malate, good growth on sucrose, and the characteristic smaller cell form.

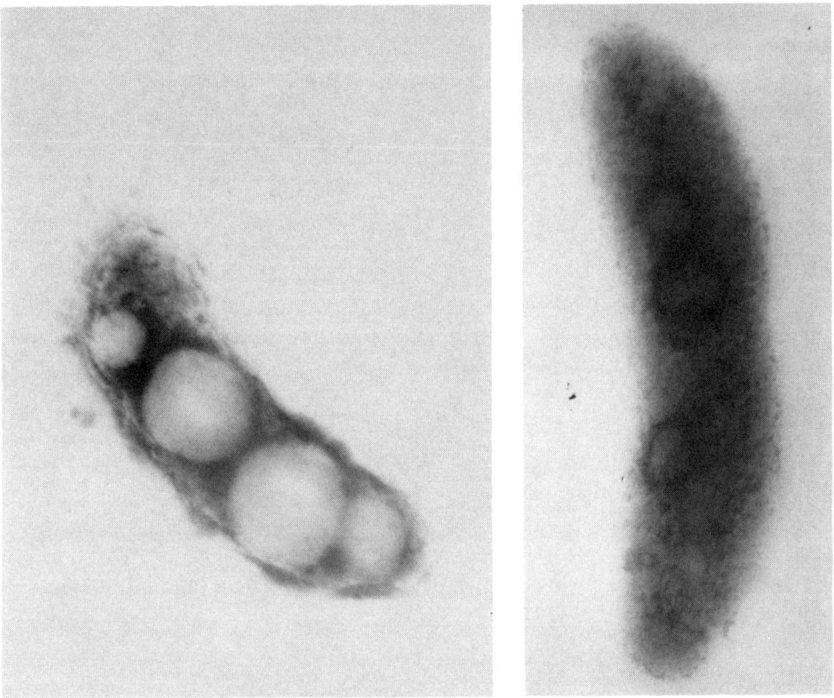

(a) (b)

Figure 2.5 Transmission electron micrographs of *A. amazonense* grown on N-free, semisolid, malate medium. (a) Cells showing PHB granules. (b) Longer, thinner cells grown with NH_4Cl (35,500×). (Çourtesy M. F. Magalhães.)

There are few additional physiological observations on this new species. The efficiency of N_2-dependent growth in semisolid media with sucrose (0.05%) was 30.9 ± 1.2 mg of N_2 (fixed) per gram of sucrose with four strains (Magalhães *et al.* 1983). Also, it was highest at late log phase when the largest proportion (11%) of fixed N was excreted possibly in the form of exopolysaccharides (Hurek 1984). In batch cultures under low p_{O2} (0.001 atm), a doubling time of 10 hours was observed in contrast to 5 to 6 hours reported for *A. brasilense* (Okon *et al.* 1977b) and *A. lipoferum* (Volpon *et al.* 1981).

The fourth species is *Azospirillum halopraeferans* Reinhold *et al.* (1987) (salt preferring). Cells are vibrioid to S shaped, and only a few helical cells are found in alkaline media. The cell width of 0.7 to 1.0 μm, can increase to 1.2 μm in alkaline media. They are motile by one polar flagellum. This species is oxidase, urease, and nitrate reductase positive and can denitrify. It does not use glucose but grows and acidifies fructose media; no growth on potato agar. It requires 0.25% NaCl for

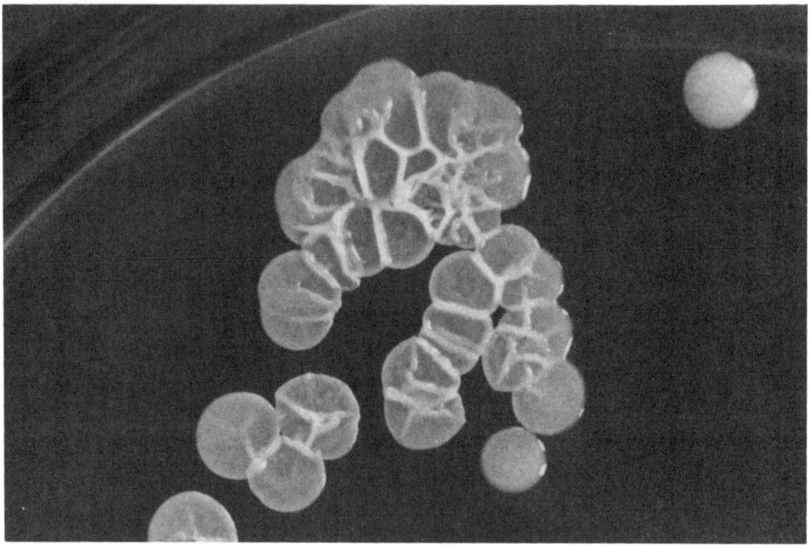

Figure 2.6 Characteristic colonies of *A. amazonense* on potato-infusion agar containing malate and sucrose.

good growth. The optimum temperature for growth is 41°C and the optimum pH range is 6.8 to 8.0. The mole percent G + C of its DNA is 68.1. This new *Azospirillum* species was found associated with Kallar grass (*Leptochloa fusca*) roots, a salt-tolerant pioneer plant common in sodic soils in Pakistan. Additional characteristics are compared in Table 2.1.

2.2 Taxonomic differentiation of *Azospirillum* from other genera

A comparison of the genus *Azospirillum* with others, according to Krieg and Döbereiner (1984), is shown in Table 2.3 and indicates the characteristics of *Azospirillum* that distinguish it from other morphologically or physiologically similar genera. In addition, several comments on taxonomy seem pertinent. An analysis of the oligonucleotide catalog obtained from the 16S rRNA of *A. brasilense* suggests that this species belongs to group Ic of the purple phototrophic bacteria as defined by Gibson *et al.* (1979). In this group, the closest known relatives of *A. brasilense* now appear to be *Aquaspirillum itersonii* and *Rhodospirillum rubrum*. Since *Azospirillum lipoferum* is closely related to *A. brasilense* as found in DNA/DNA and rRNA/DNA hybridization studies

Table 2.3 Comparison of *Azospirillum* and *Herbaspirillum* with other similar microaerobic N_2-fixing bacteria [a,b]

	Azospirillum	*Herbaspirillum*	*Aquaspirillum*	*Xanthobacter*	*Rhodospirillum*	*Rhizobium*	*Bradyrhizobium*
Cell shape							
Vibrioid	+	+	–	–	–	–	–
Helical	–	–	+	–	+	–	–
Predominant type of flagellation[c]	MP	BP	BT	–	BP	PT	MP
Growth in semisolid media—veillike pellicle	+	+	–	–	–	–	–
Nitrogenase activity	+	+	+	+	+	±	+
N_2 fixation under air	–	–	–	–	–	–	–
Growth on N_2 as sole N source	+	+	–	+	+	–	–
Anaerobic growth with NO_3^-	+[d]	–	±	–	–	–[e]	+
H_2O soluble pigments	–	–	+	–	+	–	–
Photoautrophic	–	–	–	–	+	–	–
Associated with roots	+	+	–	–	–	+	+
Mol % G + C of DNA	68–70	66–67	60–66	68–76	62–66	57–65	61–65

[a]From Baldani *et al.* 1986c, and Krieg and Döbereiner 1984.

[b]Bacteria with mol % G + C of DNA lower than 50 were not included even if they seem phenotypically similar.

[c]Symbols: MP, monopolar single flagellum; BP, one to three flagella on one or both poles; BT, bipolar tuft; PT, peritrichous; other symbols same as Table 2.1.

[d]Only *A. brasilense* and *A. lipoferum*.

[e]Except one strain of *R. leguminosarum* biovar *viceae* (Daniel *et al.* 1980).

(see below), the rRNA oligonucleotide data for *A. brasilense* almost certainly apply to *A. lipoferum* as well.

Ribosomal RNA/DNA hybridization studies by De Smedt *et al.* (1980) also support a phylogenetic relationship of *Azospirillum* to *Aquaspirillum (Rhodospirillum* has not yet been tested). The 32S rRNA cistrons of *A. brasilense* closely resemble those of *A. lipoferum* [Tm(e) is 80 to 82.5°C] and also exhibit a lower degree of similarity to those of *Aquaspirillum itersonii* [Tm(e) is 72.5°C] and *Aquaspirillum polymorphum* [Tm(e) is 71.5°C]. The studies of De Smedt *et al.* (1980) indicate that azospirilla belong to the same rRNA superfamily as *Beijerinckia, Rhizobium, Agrobacterium, Acetobacter, Gluconobacter, Xanthobacter*, and *Aquaspirillum*; members of the latter genus are presently the closest known relatives of *Azospirillum*. Although azospirilla and aquaspirilla belong in one RNA group, the inclusion of azospirilla in a separate genus seems justified both on phenotypic and phylogenetic bases. With regard to rRNA cistrons, azospirilla differ as much from aquaspirilla as azotobacters differ from *Alteromonas communis* and *Alteromonas vaga* (De Smedt *et al.* 1980).

A relationship between *Azospirillum* and *Rhodospirillum* indicated by oligonucleotide catalogs from 16S rRNA is supported by certain phenotypic evidence. Similarities concern the reversible switch off of nitrogenase promoted by NH_4^+ in *R. palustris* and the activation factor for the Fe protein of nitrogenase. This has so far been found only in *Azospirillum* and *Rhodospirillaceae* (Ludden *et al.* 1978). *Azospirillum lipoferum* and *Rhodospirillum rubrum* also share a growth requirement for biotin. Both *Azospirillum* and *R. rubrum* form a pink or red pigment when grown in the dark under aerobic conditions, but possible similarities between the pigments have not been investigated. *Azospirillum* appears not to be phototrophic, however, and *R. rubrum* also possesses bipolar tufts of flagella and appears to have cells that are thinner and more helical than those of *Azospirillum*. The mole percent G + C of the DNA of *R. rubrum* is 64 to 66, in contrast to a value of 69 to 71 for *Azospirillum*.

Certain species of the genus *Aquaspirillum* exhibit phenotypic characters in common with *Azospirillum*, namely: (a) *Aquaspirillum peregrinum, Aquaspirillum fasciculus*, and putatively, some strains of *Aquaspirillum itersonii* exhibit nitrogenase activity under microaerophilic conditions; (b) *A. itersonii, Aquaspirillum dispar*, and *Aquaspirillum psychrophilum* are denitrifiers; (c) *Aquaspirillum delicatum* occurs mainly in a vibrioid form and possesses a single polar flagellum; (d) *A. itersonii, A. peregrinum*, and *Aquaspirillum gracile* oxidizes a limited variety of sugars, and the former two species may have weak fermentative ability; (e) *A. gracile* requires biotin for growth; and (f) aquaspirilla grow best in media containing salts of organic acids. However, on the whole, the genus *Aquaspirillum* differs from *Azospirillum* by (a) having a mole percent G + C of 49 to 66, (b) usually having helical cells with bipolar tufts of flagella, (c) typically failing to oxidize or ferment carbohydrates, and (d) not being associated with plants and living mainly in an aquatic habitat.

There is little evidence, genetic or phenotypic, that azospirilla are closely related to the genus *Azotobacter*, although *Azotobacter paspali* is, like azospirilla, associated with the roots of plants. Azospirilla, however, lack the respiratory protection against oxygen inactivation of nitrogenase that is found in *Azotobacter chroococcum* (Robson and Postgate 1980), and the ultrastructure of the C-forms of azospirilla have little similarity to the cysts of azotobacters. Moreover, azotobacters are not vibrioid in shape and are not motile by virtue of a single polar flagellum in liquid media.

Although the mole percent G + C of the DNA of the N_2-fixing genus *Derxia* is similar to that of azospirilla, *Derxia* differs from azospirilla in (a) being a straight rod, and (b) forming very large, distinctive, very dense colonies that enable them to fix N_2 under air. The genus *Xanthobacter* resembles azospirilla in being a microaerophilic diazotroph and in having a mole percent G + C of 69 to 70. However, the cells are smaller and under certain conditions exhibit branching, are nonmotile, and form a yellow pigment.

The N_2-fixing genus *Rhizobium* resembles azospirilla in being plant associated, but the host–bacterium interaction is far more highly specialized for symbiotic N_2 fixation than is the case with azospirilla.

The vibrioid shape, occurrence of lateral flagella when grown on solid media, and occurrence of some fermentative ability suggest a relationship between azospirilla and the genus *Vibrio*. However, the mole percent G + C of the DNA of *Vibrio* is 38 to 51, well below that of azospirilla.

2.3 Physiology and biochemistry

Physiological and biochemical studies on *Azospirillum* spp. have covered nutritional requirements, O_2 protection, hydrogenase–nitrogenase interactions, carbohydrate metabolism, hydrogen and nitrogen metabolism, aerotaxis and chemotaxis, autotrophy and methylotrophy, and nitrogenase properties. *Azospirillum brasilense* and *A. lipoferum* are the best studied among the azospirilla. *Azospirillum amazonense*, although a more recently described organism, is well characterized in terms of carbohydrate metabolism, O_2 protection, and nitrogenase properties and regulation. Few data are available on *A. halopraeferans*.

Carbon metabolism The four *Azospirillum* species differ markedly with respect to carbon source utilization. Organic acids are the preferred carbon sources for all species (Magalhães *et al.* 1983; Okon *et al.* 1976a; Reinhold *et al.* 1985a; Tarrand *et al.* 1978). The pattern of carbohydrate utilization, however, is sufficiently species specific to allow for their taxonomic intragenus identification (Table 2.4). *Azospirillum amazonense* is remarkable for its ability to grow and fix N_2 in media containing disaccharides (Magalhães *et al.* 1983; Martinez-Drets *et al.* 1984), which are not used by the other species. Apparently, any of the metabolizable C-substrates also support nitrogen fixation and diazotrophic growth. The degree of efficiency of any given C-substrate in supporting acetylene reduction activity during growth in N-free semisolid media varies with the authors, and disputes as to which is the best substrate seem irrelevant. Growth media and experimental conditions could be responsible for the observed variations. Most of the studies were carried out using buffers of low molarity, and the pH values of the media increased rapidly during growth on organic acids. Few systematic studies comparing growth parameters for different substrates under well-defined conditions are available.

For additional information on carbon sources for growth of *Azospirillum*, see Das and Mishra (1983), Döbereiner and Day (1976), Haahtela *et al.* (1983), Halsall *et al.* (1985), Martinez-Drets *et al.* (1984, 1985), Okon *et al.* (1976a), Pedrosa *et al.* (1980), Sampaio *et al.* (1982), Tarrand *et al.* (1978), and Westby *et al.* (1983).

The metabolism of sucrose, D-glucose, D-fructose, L-arabinose, and D-gluconate have been determined in *Azospirillum* spp. (Goebel and Krieg 1984; Loh *et al.* 1984; Martinez-Drets *et al.* 1984, 1985; Novick and Tyler 1982; Westby *et al.* 1983). *Azospirillum brasilense* (strain Sp7) and *A. lipoferum* (strain Sp59b) catabolize D-fructose via the glycolytic pathway or Embden–Meyerhof–Parnas (EMP) pathway (Goebel and Krieg 1984; Martinez-Drets *et al.* 1984). Sucrose-grown *A. amazonense* (Y1) lacks this pathway and catabolizes D-glucose and D-fructose, produced by the action of β-fructo furanosidase on sucrose exclusively via the Entner–Doudoroff (ED) pathway (Martinez-Drets et al. 1985), which is also present in *A. lipoferum* (Goebel and Krieg 1984; Martinez-Drets *et al.* 1984). D-Fructose-grown *A. lipoferum* has a low level of a key enzyme of the ED pathway, namely, 6-phosphogluconate dehydratase, but contains 2-keto-3-deoxy-6-phosphogluconate aldolase (Goebel and Krieg 1984; Martinez-Drets *et al.* 1984; Westby *et al.* 1983). In addition *A. brasilense* has very low (Westby *et al.* 1983) or undetectable levels of NAD(P)-glucose 6-P-dehydrogenase (Goebel and Krieg 1984; Martinez-Drets *et al.* 1984), an enzyme present in high levels in *A. lipoferum* and *A. amazonense* (Goebel and Krieg 1984; Martinez-Drets *et al.* 1984, 1985) and which is required for 6-phosphogluconate synthe-

Table 2.4 Carbohydrate metabolism in *Azospirillum* spp.[a,b]

	Embden–Meyerhof–Parnas pathway				Entner–Doudoroff pathway				Hexose monophosphate pathway			
Species	Fru	Glu	Suc	Gluc	Fru	Glu	Suc	Gluc	Fru	Glu	Suc	Gluc
A. brasilense (Sp7, Cd)	+	a	a	+(1)	−(2)	a	a	+	−(3)	a	a	−(3)
A. lipoferum (Sp59b, Col5)	+	+	b	nd	+	+	b	nd	−(3)	−(3)	b	nd
A. amazonense (Y1, Y2, Y6, 591, 585, M132)	nd	nd	−(4)	c	nd	nd	+(5)	c	nd	nd	−(3)	c

[a]Abbreviations and symbols: Fru, D-fructose; Glu, D-glucose; Suc, sucrose; Gluc, D-gluconate; +, all enzymes present; −, key enzymes absent; nd, not determined; a, *A. brasilense* does not grow on D-glucose or sucrose; b, *A. lipoferum* does not grow on sucrose; c, *A. amazonense* does not grow on D-gluconate; (1), lacks 6-phosphofructokinase when grown on D-gluconate, glycerol, lactate, and malate (Westby *et al.* 1983); (2), lacks glucose 6-phosphate dehydrogenase and 6-phosphogluconate dehydratase, but has 2-keto-3-deoxy-D-gluconate 6-phosphate aldolase when grown on D-fructose (Goebel and Krieg 1984; Martinez-Drets *et al.* 1984); (3), lacks the key enzyme of the oxidative branch of the HMP pathway, $NADP^+$-6-phosphogluconate dehydrogenase; there are no reports on the enzymes of the isomerization branch, e.g., transaldolase and transketolase; (4), lacks 6-phosphofructokinase, 1-phosphofructokinase and fructose 1,6-biphosphate aldolase; (5), sucrose is first hydrolyzed by β-fructofuranosidase.

[b]Data from Goebel and Krieg 1984, Martinez-Drets *et al.* 1984, 1985, and Westby *et al.* 1983.

sis. All three species lack the key enzyme of the oxidative branch of the hexose monophosphate pathway (HMP), NADP$^+$-6-phosphogluconate dehydrogenase (Goebel and Krieg 1984; Martinez-Drets *et al.* 1984, 1985; Westby *et al.* 1983). The enzymes of the nonoxidative isomerization branch of the HMP pathway, transaldolase and transketolase, have not been measured in any of the species, but must be present to ensure a pentose supply for nucleic acid synthesis.

In *A. brasilense*, the glycolytic substrate D-fructose, effectively induced a key enzyme of the catabolic EMP pathway, 6-phosphofructokinase, while repressing that of the anabolic EMP pathway (gluconeogenesis), D-fructose 2,6-bisphosphatase (Goebel and Krieg 1984; Martinez-Drets *et al.* 1984; Westby *et al.* 1983). The opposite situation was true for the gluconeogenic substrates glycerol, lactate, malate, and D-gluconate. This kind of regulatory control has not yet been determined for *A. lipoferum*. To participate in either of the above catabolic pathways, the hexose must be first phosphorylated at the 6-position. Glucokinase and hexokinase in *A. lipoferum* and fructokinase and glucokinase in *A. amazonense* carry out the phosphorylation of both D-glucose and D-fructose. In *A. brasilense*, however, only glucokinase, which is specific for D-glucose, is present and the phosphorylation of D-fructose occurs in the 1-position via an inducible phosphoenolpyruvate–phosphotransferase system (PEP–PTS) (Goebel and Krieg 1984; Gupta and Ghosh 1984). This system is also operative in *A. lipoferum* (Goebel and Krieg 1984). The follow-up enzyme associated with this phosphorylation system, 1-phosphofructokinase, is present in large amounts in *A. brasilense* and *A. lipoferum* (Goebel and Krieg 1984; Martinez-Drets *et al.* 1984). It is lacking in sucrose-grown *A. amazonense* (Martinez-Drets *et al.* 1985), thus suggesting the absence of the D-fructose PEP–PTS also.

Studies using ^{15}C as a tracer have shown that *A. brasilense* is unable to take up D-glucose (Table 2.5), explaining its inability to metabolize this sugar despite having all the required enzymes (Goebel and Krieg 1984; Loh *et al.* 1984; Martinez-Drets *et al.* 1984). *Azospirillum brasilense* is thus cryptic for D-glucose metabolism. Respirometric studies have indeed shown that a cell-free extract of D-fructose-grown *A. brasilense* strain Sp7 oxidizes D-glucose (Das and Mishra 1983). Membrane leakiness may explain why few (~9%) *A. brasilense* strains T60, 67, F6, and 34 (Tarrand *et al.* 1978) and strains K67 and Kr77 (Franche and Elmerich 1981) grew poorly on D-glucose. Attempts to demonstrate a glucose PEP–PTS in *A. brasilense* and *A. lipoferum* were unsuccessful (Goebel and Krieg 1984). *Azospirillum brasilense* and *A. lipoferum* metabolize D-gluconate via the ED pathway (Martinez-Drets *et al.* 1984; Westby *et al.* 1983). The enzymes gluconokinase and 6-phosphogluconate dehydratase are inducible by gluconate, while 2-keto-3-deoxy-

Table 2.5 Uptake of ^{14}C sources by *Azospirillum* spp.[a]

	D-Fructose	D-Glucose	Sucrose	Mannitol	α-Keto glutarate
A. brasilense[b]	+	−	−	nd	−
A. lipoferum[c]	+	+	−	nd	+
A. amazonense[c,d]	nd	+	+	−	−

[a]Symbols: +, positive; −, negative; nd, not determined.

[b]Cells grown in minimal media with D-fructose supplemented with D-glucose, sucrose, or α-ketoglutarate (Martinez-Drets *et al.* 1984), or in complex media with D-glucose (Goebel and Krieg 1984).

[c]Cells grown in minimal media with D-glucose, sucrose, mannitol, or α-ketoglutarate.

[d]*A. amazonense* does not grow on D-gluconate; this is probably due to the lack of a transport system (Martinez-Drets *et al.* 1985).

6-phosphogluconate aldolase is constitutive in *A. brasilense* grown in malate, lactate, glycerol, and D-fructose (Martinez-Drets *et al.* 1984; Westby *et al.* 1983). In *A. lipoferum*, the three enzymes are constitutive for D-glucose- and D-fructose-grown cells (Goebel and Krieg 1984; Martinez-Drets *et al.* 1984). Although sucrose-grown *A. amazonense* has the three above-mentioned enzyme activities, it is unable to grow on D-gluconate probably because of its lack of a gluconate transport system (Martinez-Drets *et al.* 1985). The absence of NAD(P)-D-glucose-6-phosphate dehydrogenase activity in *A. brasilense* Sp7 and Cd, independent of the carbon source for growth (Goebel and Krieg 1984; Martinez-Drets *et al.* 1984; Westby *et al.* 1983), prevents the formation of 6-phosphogluconate and, consequently, the catabolism of D-fructose via the ED pathway. The presence of D-glucose-6-phosphate dehydrogenase and the enzymes of the ED pathway together with the enzymes of the EMP pathway in D-glucose- or D-fructose-grown *A. lipoferum* suggests that both pathways may be operative simultaneously. In contrast, *A. amazonense* will catabolize D-glucose and D-fructose exclusively by the ED pathway (Martinez-Drets *et al.* 1984, 1985). *A. brasilense* will also metabolize 2-ketogluconate (Westby *et al.* 1983). Another carbon source utilized by most strains, L-arabinose, is metabolized by *A. brasilense* via an inducible oxidative pathway with no phosphorylated intermediates (Novick and Tyler 1982). In this pathway, L-arabinose is converted into α-ketoglutarate in a manner identical to that described for fast-growing *Rhizobium* (Duncan 1979). The enzymes of this pathway, L-arabinose dehydrogenase, L-arabinolactonase, L-arabinate dehydratase, 2-keto-3-deoxy-L-arabinate dehydratase, and 2-keto-glutaratesemialdehyde dehydrogenase are present in L-arabinose-grown *A. brasilense* Sp7 (Novick and Tyler 1982) and L-arabinose dehydrogenase has been partially purified (Novick and Tyler 1983). The 2-keto-glutarate then enters the tricarboxylic acid (Krebs) cycle.

Glycerol is fed into the EMP pathway at the level of dihydroxyacetone phosphate and involves glycerokinase and 3-phosphate dehydrogenase activities. The former enzyme is induced by growth of *A. brasilense* in glycerol, but can also be found in cells grown on lactate, malate, and gluconate (Westby *et al.* 1983).

Azospirillum brasilense and *A. lipoferum* are unable to utilize starch, cellulose, and carboxy-methyl cellulose (Halsall *et al.* 1985; Tarrand *et al.* 1978; Wong *et al.* 1980). *Azospirillum lipoferum* and *A. brasilense* strains are, however, capable of utilizing straw and xylan (a major component of straw) for diazotrophic growth with ^{15}N incorporation (see Table 2.11) (Halsall *et al.* 1985). Extensive degradation of xylan and straw occurred with 35 and 12.5%, respectively, of the carbon present converted to CO_2 after 14 days (Halsall *et al.* 1985). Extracellular xylanase activity was found in *Azospirillum* strains capable of straw and xylan utilization (Halsall and Gibson 1985).

The production of polygalacturonic acid transeliminase (pectic lyase) by several *Azospirillum* strains (Tien *et al.* 1981) suggests that pectins, another significant component of plant residues, may also be utilized. Transeliminase production was stimulated by polygalacturonic acid, and the enzyme appears to be bound to the cell wall. Growth in the presence of amino acid rather than ammonium salts increased the ratio of free to bound enzyme. The enzyme acted preferentially on polygalacturonic acid, although it also cleaved pectin. The optimum pH for activity was 8.0, and $CaCl_2$ activated the enzyme (Tien *et al.* 1981).

Poly-β-hydroxybutyrate (PHB) is a carbon and energy reserve polymer widely distributed throughout the microbial world. High amounts of PHB (39 to 40%) accumulate in *A. brasilense* and *A. lipoferum* grown diazotrophically (Okon *et al.* 1976b; Tal and Okon 1985; Volpon *et al.* 1981). NH_4^+-grown cells, in contrast, have low levels ($<1\%$) of PHB (Okon *et al.* 1976b). Accumulation of PHB, however, occurred in *A. brasilense* strain Cd grown chemostatically in medium containing NH_4^+ under O_2 limitation (Nur *et al.* 1982a, b). Thus similarly to what occurs in *Azotobacter beijerinckii*, O_2 limitation, rather than N-limitation is a determining factor in PHB accumulation in *A. brasilense* (Tal and Okon 1985). Indeed, an extreme case is that of *A. brasilense* grown anaerobically in medium containing NO_2^- as a N source and as a respiratory electron acceptor, which accumulated large amounts of PHB only after exhaustion of NO_2^- (Zimmer *et al.* 1984). In contrast, PHB accumulated during anaerobic growth in the presence of NH_4^+ and N_2O (Stephan *et al.* 1984) suggesting that N_2O is a poor respiratory electron acceptor for *A. brasilense*. Extensive studies by Tal and Okon (1985) identified several carbon- and energy-dependent physiological functions supported by PHB in *A. brasilense*. PHB supported growth and respiration, nitrogen fixation, an increased survival, resistance to desiccation, resistance to osmotic shock and pressure, and resistance to ultraviolet radiation. The high PHB content of washed, C-free cell suspensions of *A. brasilense* supported high nitrogenase activity and made it difficult to demonstrate H_2-dependent nitrogenase activity (Pedrosa *et al.* 1982).

Enzymatic studies have shown that the levels of β-hydroxybutyric acid dehydrogenase (NAD^+-dependent) of *A. brasilense* are high in diazotrophically grown cells (Okon *et al.* 1976b).

The tricarboxylic acid (Krebs) cycle has been shown to be operative in all *Azospirillum* spp., as inferred from the utilization of cycle intermediates as sole carbon sources for growth (Okon *et al.* 1976a), radiorespirometric studies (Loh *et al.* 1976a), and enzymatic studies (Goebel and Krieg 1984; Martinez-Drets *et al.* 1984, 1985; Westby *et al.* 1983). Also present are the key enzymes of the glyoxylic acid cycle, isocitrate lyase and malate synthase (Goebel and Krieg 1984; Martinez-Drets *et al.* 1984, 1985), essential for growth on two-carbon substrates such as ethanol and acetate. Other key enzymes also present in *A. brasilense* and in *A. lipoferum* are lactate dehydrogenase, the pyruvate dehydrogenase complex, and malic enzyme (Martinez-Drets *et al.* 1984; Westby *et al.* 1983), which are essential for growth on lactate, pyruvate, and tricarboxylic acid cycle intermediates. *Azospirillum amazonense* and *A. halopraeferans* must also have these enzymes.

Methylotrophic growth was observed in *A. brasilense* Sp7 and *A. lipoferum* Sp208, which showed significant growth on C-1 compounds such as methane, methanol, and formate (Sampaio *et al.* 1981). Nevertheless, attempts to demonstrate $^{14}CH_4$ incorporation into *A. brasilense* and *A. lipoferum* biomass have failed (Sampaio *et al.* 1982). These reports of growth of *Azospirillum* spp. on C-1 compounds have not been confirmed by other workers. Hydrogen-dependent autotrophic growth was first observed in *A. lipoferum* strains RG20 and Sp208 which are capable of H_2-dependent carbon dioxide fixation (Pedrosa *et al.* 1980). The presence of H_2-uptake hydrogenase and ribulose 1,5-bisphosphate carboxylase activity in H_2- plus CO_2-grown *A. lipoferum* strain Sp208 (Sampaio *et al.* 1981) confirmed their potential for autotrophic growth. These observations were extended to other *A. lipoferum* strains (Sp59b, A3a, RG20a, RG6xx, and Br17) (Malik and Schlegel 1981). *A. lipoferum* strains were also shown to be capable of limited diazotrophic growth under an atmosphere of H_2, CO_2, N_2, and O_2 (Malik and Schlegel 1981; Sampaio *et al.* 1981) and simultaneous derepression of hydrogenase, D-ribulose-1,5-bisphosphate carboxylase, and nitrogenase was observed (Sampaio *et al.* 1981). All *A. brasilense* strains tested so far failed to grow autotrophically (Gowda and Watanabe 1983; Malik and

Schlegel 1981; Pedrosa *et al.* 1980), and there are no reports on autotrophic growth of *A. amazonense* or *A. halopraeferans.*

Respiratory chain and oxidative phosphorylation *Azospirillum* spp. are aerobes displaying oxidative metabolism with practically no fermentative ability (Tarrand *et al.* 1978). Organic acids such as malate, lactate, pyruvate, or succinate which support growth also support high rates of O_2 uptake by intact cell suspensions (Okon *et al.* 1976b). Sugars, with the exception of fructose (Das and Misra 1983), are poor respiratory substrates (Okon *et al.* 1976b). In cell-free extracts, however, glucose, galactose, sucrose, and fructose showed O_2 uptake rates similar to those of malate and lactate (Das and Mishra 1983). The components of the respiratory chain of *A. brasilense* have been determined spectroscopically in cells grown on N_2 or NH_4^+ under high and low dissolved O_2 concentrations (Burris *et al.* 1978; Okon *et al.* 1983; Reiner *et al.* 1985). Cell-free extracts of N_2-grown *A. brasilense* showed absorption peaks corresponding to cytochromes *b* (α, 559 nm), *c* (α, 553 nm), and aa_3 (α_{max}, 600 nm) (Burris *et al.* 1978). Membrane preparations of *A. brasilense* grown in NH_4^+ under low and high dissolved O_2 concentrations had cytochrome *b* (λ_{max} and α_{max}, 428 and 560 nm, respectively) and cytochrome *c* (α_{max}, 551 nm, probably fused peaks). An additional peak of α_{max} 603 nm identified as cytochrome aa_3 was found in membranes from cells grown at high dissolved O_2 concentrations, but only in trace amounts in cells grown at low dissolved O_2 concentrations (Okon *et al.* 1983). A soluble cytochrome *c* is always present in *A. brasilense* grown under low and high O_2 concentrations (Burris *et al.* 1978; Okon *et al.* 1983). A light membrane fraction is found in cells grown at high dissolved oxygen concentrations and contains cytochrome *b* and aa_3, but lacks cytochrome *c* (Okon *et al.* 1983).

The entire assembly of components required for succinate and NADH oxidation was found associated with membranes in *A. brasilense* (Okon *et al.* 1983). Respirometric studies showed that CO inhibited succinate oxidase activity in membranes from cells grown at high O_2 concentrations, but stimulated that from cells grown at low concentrations. This suggests the presence, in cells grown under low oxygen concentrations, of a terminal oxidase other than cytochrome aa_3 (Okon *et al.* 1983). The effect of CN^- concentrations on succinate and NADH oxidase activities in membranes from *A. brasilense* suggests the presence of a branched respiratory electron chain in both types of membranes (Okon *et al.* 1983). Further kinetic studies with membranes from cells grown at high dissolved O_2 concentrations indicates two pathways for the oxidation of NADH with two terminal oxidases: a low affinity oxidase (K_s, apparent dissociation constant: 28 μM O_2) completely inhibitable by

0.1 m*M* CN^-, which only slightly affected the high affinity oxidase (K_s, 1.5 to 4.2 μM O_2) (Okon *et al.* 1983).

Kinetic studies with N_2-fixing cell suspensions of *A. brasilense* grown chemostatically from 0 to 11 μM dissolved O_2, were carried out at very low dissolved O_2 concentrations using leghemoglobin or myoglobin as O_2 carriers (Bergersen and Turner 1980). A single very high affinity terminal oxidase (K_s, 0.006 μM O_2) was found in all N_2-fixing cultures. An additional terminal oxidase (K_s, 0.5 μM O_2) was found in cells grown at 11 μM O_2 and was regarded as "protective oxidase" (Bergersen and Turner 1980). Therefore, three terminal oxidases are present in *A. brasilense* as indicated by their affinity constants. Additional studies have shown that in addition to cytochrome aa_3, *A. brasilense* has the terminal oxidases cytochrome *o* and cytochrome *d* (Reiner *et al.* 1985).

Inhibition studies led Okon *et al.* (1983) to suggest that NADH and succinate oxidation by *A. brasilense* membranes proceeds to oxygen through a branched pathway, with a branch point situated on the substrate side of the antimycin A block (cyt *b*560 → cyt *c*551). According to this model, cytochrome *c* only donates or accepts electrons after the site of antimycin A inhibition on the pathway that is sensitive to low KCN concentrations, assuming that this pathway terminates in cytochrome *aa*3 (Okon *et al.* 1983). The other arm of the branched chain is assumed to be blocked by a high concentration of KCN and is the only pathway for succinate oxidation (Okon *et al.* 1983).

In membranes from cells grown at low dissolved O_2 concentrations both NADH and succinate oxidase activities were completely inhibited by antimycin A, and succinate oxidase activity inhibition of KCN showed a biphasic pattern (Okon *et al.* 1983). This suggests that two branches for succinate oxidation are present in these membranes.

The data, however, do not exclude the possibility that parallel, independent (instead of branched) respiratory chains are responsible for the oxidation of specific substrates, as seems to be the case for the simultaneous oxidation of H_2 and organic acids by *A. brasilense* (Pedrosa *et al.* 1982).

There are no direct estimates of oxidative phosphorylation by *Azospirillum* spp., although relative efficiencies for some substrates and electron acceptors are available. For instance, H_2 is a less efficient respiratory substrate than organic acids as evident at suboptimal O_2 concentrations (Pedrosa *et al.* 1982). On the other hand, O_2 and NO_2^- are three times more efficient as terminal electron acceptors than NO_3^- in increasing molar growth yields on malate (Danneberg *et al.* 1985). In the report by Danneberg *et al.* the contribution of PHB to biomass was not stated, and the data should therefore be taken with caution. Preliminary results suggest that cytochrome *o* appears to be the primary O_2

receptor in aerotaxis, although the participation of cytochromes *b* and *c* has not been excluded (Reiner *et al.* 1985).

Hydrogen metabolism Hydrogen metabolism by bacteria and cyanobacteria involves the participation of three distinct enzymes (for reviews, see Adams *et al.* 1981; Evans *et al.* 1981; Robson and Postgate 1980). Reversible hydrogenases are found in anaerobes and have the main functions of removal of excess reducing power in heterotrophically grown cells, and of H_2 oxidation to produce ATP via anaerobic oxidative phosphorylation during H_2-dependent growth (Adams *et al.* 1981). Uptake hydrogenases are unidirectional, irreversible in action *in vivo*, and are involved in utilizing H_2 as an energy source for metabolism. They are found in aerobes, facultative anaerobes, and anaerobes, acting similarly to the oxidative function of the reversible hydrogenases. Electrons derived from H_2 are transferred to terminal electron acceptors by the respiratory chain, and ATP is generated in the process. Dixon (1972) postulated three physiological roles for uptake hydrogenases of diazotrophs: (a) to remove inhibitory hydrogen from the vicinities of nitrogenase; (b) to recycle H_2 produced by nitrogenase activity; and (c) to protect nitrogenase from O_2 inactivation by enhancement of O_2 consumption, a H_2-dependent kind of respiratory protection. Studies with *Azotobacter chroococcum* lend support to the last two postulates while contradicting the former and also demonstrate that H_2 can provide reducing power for N_2 reduction (Walker and Yates 1978).

The third enzyme involved in H_2 metabolism is nitrogenase, which has an ATP-dependent H_2 evolution activity. This activity is highest in the absence of reducible substrates and minimal (ratio of H_2 produced to N_2 reduced is 1.0) in the presence of saturating N_2, or even at 50 atm N_2 (Simpson and Burris 1984). The H_2:N_2 ratio, which reflects the efficiency of energy utilization by nitrogenase in reducing N_2, increases at low temperatures, at high MoFe protein/Fe protein ratios, and at low ATP/ADP ratios (Thorneley and Eady 1977). The effect of low ATP/ADP ratios is related to inhibition of Fe protein by ADP, thus increasing the functional MoFe/Fe protein ratio (Thorneley and Eady 1977). Factors such as pH, temperature (Andersen and Shanmugam 1977), oxygen, carbon substrates, and the limiting nutrient (Mo, S, Fe) (Yates *et al.* 1981), which directly or indirectly affect the functional MoFe/Fe protein ratio or the ATP/ADP ratio, cause variations in the *in vivo* ratio of H_2 evolved/N_2 reduced by cultures of *Klebsiella pneumoniae* and *Azotobacter chroococcum*. Substantial amounts (25 to 60%) of the energy supplied *in vivo* to nitrogenase are used to produce H_2 and could be recovered through the action of an uptake hydrogenase (Evans *et al.* 1981).

Azospirillum brasilense strain Sp7 and *A. lipoferum* strains Br17 and Sp208b contain H_2 uptake hydrogenase activity sufficient to recycle all hydrogen produced by nitrogenase (Berlier and Lespinat 1980; Chan *et al.* 1980; Pedrosa *et al.* 1981; Stephan *et al.* 1981; Volpon *et al.* 1981). Oxygen and methylene blue, nitrous oxide, and nitrate all serve as electron acceptors for hydrogenase of *A. brasilense* grown anaerobically on nitrate or nitrous oxide (Tibelius and Knowles 1984a). The enzyme activity is inhibited by CO, O_2, NO, NO_2, and C_2H_2 (Berlier and Lespinat 1980; Chan *et al.* 1980; Pedrosa *et al.* 1981; Stephan *et al.* 1981; Tibelius and Knowles 1984b). Hydrogenase synthesis occurs equally well in the presence or absence of NH_4^+ ions as long as the dissolved O_2 concentrations are kept low (<9 μM) (Tibelius and Knowles 1983; F. O. Pedrosa, unpublished data). Neither H_2 nor nitrogenase derepression are required for hydrogenase synthesis in *A. brasilense*, but low carbon substrates or electron donor levels seem to increase hydrogenase activity (Pedrosa *et al.* 1981; Tibelius and Knowles 1983). However, H_2 increases hydrogenase stability in stationary cultures of *A. brasilense* (Tibelius and Knowles 1983). In *A. brasilense* and *A. lipoferum* cultures sparged with a fixed O_2 concentration (0.003 atm) (Stephan *et al.* 1981; Volpon *et al.* 1981) or anaerobically with N_2O or NO_3^- (Tibelius and Knowles 1984a), hydrogenase activity increased significantly at the end of the log phase, a phenomenon possibly related to carbon substrate limitation. Increased synthesis of hydrogenase under C-limitation has already been shown (Pedrosa *et al.* 1981).

Evolution of H_2 in N_2-grown *A. brasilense* occurred only when CO or C_2H_2 (Berlier and Lespinat 1980; Chan *et al.* 1980; Stephan *et al.* 1981) was present and followed the same pattern of response to O_2 as nitrogenase activity (Chan *et al.* 1980). In addition, growth of *A. brasilense* and *A. lipoferum* in the presence of chelating agents such as orthophenanthroline (20 μM) or EDTA (250 μM), which do not inhibit preformed hydrogenase nor nitrogenase activity, caused maximum H_2 evolution (Pedrosa and Yates 1983a). This effect was specifically reversed by Ni^{2+}, which eliminated H_2 evolution and increased hydrogenase activity without affecting that of nitrogenase. Nickel ions are thus required for hydrogenase synthesis and activity, and this requirement could not be fulfilled by Fe^{2+}, Ca^{2+}, Mn^{2+}, Zn^{2+}, or Cu^{2+} (Pedrosa and Yates 1983a). The H_2 uptake hydrogenase of *Azospirillum* spp. is probably of the NiFe type similar to that of sulfate-reducing bacteria (Odom and Peck 1984). Berlier and Lespinat (1980) claim the presence of reversible hydrogenase activity in *A. brasilense*, which would evolve H_2 or H^2H under anaerobic conditions. Pedrosa *et al.* (1982) observed H^3H uptake by *A. brasilense* Sp7 anaerobically in the presence of sodium dithionite to scavenge O_2 traces. This anaerobic activity was, however, less than 5% of that in the presence of optimum O_2 concentrations, and it may represent "exchange" activity or possibly electron transfer to endogenous acceptors such as β-ketobutyrate. Increase in β-hydroxybutyrate dehydrogenase and poly-β-hydroxybutyrate accumulation under low O_2 concentrations (Nur *et al.* 1982b; Okon *et al.* 1976b) suggest that this may be true. Indeed, *A. brasilense* grown anaerobically in continuous culture under NO_2^- limitation accumulated large amounts of PHB (Zimmer *et al.* 1984). Anaerobic H_2 uptake occurred in *A. brasilense* strain Sp7 grown anaerobically in the presence of N_2O or NO_3^-, and had higher activity with O_2 as the terminal electron acceptor than with N_2O or NO_3^- (Tibelius and Knowles 1984a).

The uptake hydrogenase activity of N_2-fixing *A. brasilense* is capable of supporting H_2-dependent acetylene reduction by cells starved of carbon metabolites, although carbon substrates do not inhibit H_2-dependent respiration (Pedrosa *et al.* 1982). Hydrogen oxidation cannot support respiratory protection of nitrogenase because the O_2-dependent H_2 uptake system is itself extremely sensitive to inhibition by O_2 (Pedrosa *et al.* 1982). Tibelius and Knowles (1983) disputed the latter conclusion since they found that the optimum p_{O_2} for the H_2 uptake system was significantly higher than that for nitrogenase activity. In their experiments, however, the effect of p_{O_2} on H_2 uptake and on C_2H_2 reduction activities was determined at different shaking rates, namely, 250 and 300 rpm, respectively. At lower speeds the rate of O_2 diffusion is decreased and consequently higher O_2 optima are found. Furthermore, nitrogenase and hydrogenase activities derepressed simultaneously during batch culture growth of *A. brasilense* in glutamate under air (F. O. Pedrosa and M. G. Yates, unpublished data) suggesting similar O_2 sensitivity. Derepression of nitrogenase was also observed in media with several other amino acids (Das and Mishra 1982b; Gallori and Bazzicalupo 1985; Hartmann *et al.* 1985).

In *A. brasilense* H_2-dependent respiration was inhibited by KCN (K_i, 20 μM) whereas malate-dependent respiration was unaffected by KCN at p_{O_2} 0.005 atm (Pedrosa *et al.* 1982). Malate-dependent respiration, however, was inhibited by KCN (K_i, apparent inhibitor dissociation constant; 32 μM O_2) under air. This and the noninhibitory effect of malate on H_2-dependent respiration suggest that H_2 and carbon-linked respiratory chains are independent with no common electron carriers as in a branched electron transport chain (Pedrosa *et al.* 1982).

Hydrogen partly inhibits nitrogenase activity at suboptimal O_2 concentrations possibly because H_2-dependent respiration has a greater affinity for O_2, but is less efficiently coupled to ATP production than carbon-dependent respiration (Pedrosa *et al.* 1981, 1982).

Nitrogen metabolism *Azospirillum* spp. may perform all steps of the nitrogen cycle except nitrification. *Azospirillum brasilense* and *A.*

lipoferum are thus capable of (a) nitrogen fixation, (b) assimilatory nitrate reduction, (c) NH_4^+ assimilation, and (d) denitrification or dissimilatory nitrate reduction.

Nitrogen fixation Diazotrophic growth of azospirilla occurs only at very low dissolved O_2 concentrations (Day and Döbereiner 1976; Magalhães *et al.* 1983; Okon *et al.* 1976a). Growth on NH_4^+ is not affected by air (Day and Döbereiner 1976), although the organisms tend to migrate to regions of lower p_{O_2} (Barak *et al.* 1982). Aerobic nitrogen-fixing organisms have evolved strategies or mechanisms to cope with oxygen, an inactivator and repressor of nitrogenase activity and synthesis (Bergersen 1984; Gallon 1981; Postgate 1982; Robson and Postgate 1980). These nitrogenase protection mechanisms against O_2 include the following:

1. Respiratory protection. Azotobacters respond to increased O_2 concentrations by increasing their respiratory rate and, in consequence, removing excess O_2.
2. Conformational protection. Cultures cease to fix nitrogen when briefly exposed to high O_2 concentrations, but resume fixation immediately after removal of excess O_2. In this switch on/off process mediated by oxygen in *Azotobacter*, one or both of the following may occur: (a) conformational change in nitrogenase which renders it oxygen tolerant, with an additional FeS protein involved, or (b) discharge of nitrogenase electron donors through autooxidation.
3. Compartmentalization. The N_2-fixing system is restricted to a compartment protected against oxygen. Aerobic cyanobacteria restrict their sites of nitrogen fixation to specialized organelles, the heterocysts that have thick cell walls and do not evolve photosynthetic O_2. The nodule is a specialized plant organelle developed in the legume–*Rhizobium* symbiosis. The nitrogen-fixing *Rhizobium* bacteroids may be protected against O_2 inactivation by leghemoglobin. This hemeprotein efficiently binds and transports O_2 to the bacteroids while maintaining the free dissolved O_2 concentration at a minimum.
4. Enzymatic protection. Toxic O_2 reduction metabolites such as superoxide ion (O_2^-), peroxide (H_2O_2), hydroxyl radical ($HO^\cdot$), and singlet oxygen are probably damaging to MoFe and Fe proteins. Superoxide dismutase, peroxidase, and catalase may afford additional protection to nitrogenase of aerobic diazotrophs by removing these metabolites.

5. Chemical protection. Carotenoids are thought to give additional protection by quenching singlet oxygen produced by the spontaneous dismutation of superoxide and possibly oxygen radicals (Krinsky 1979).
6. Restriction of O_2 diffusion. (a) Cellular aggregation: respiration of the outer cell layer may prevent O_2 diffusion to the central N_2-fixing cells; (b) extracellular gum production: in some bacteria heavy gum production may restrict O_2 diffusion.
7. Avoidance of oxygen. Anaerobic diazotrophs fix nitrogen only in the absence of O_2, a prerequisite for cell survival. Thus, anaerobes show passive avoidance of O_2. Aerobic diazotrophs may show active avoidance expressed by aerotaxis, an active cell motion toward regions of optimum O_2 concentration.

Limited respiratory protection, aerotaxis, conformational protection, and the presence of carotenoids have been suggested to protect nitrogenase against O_2 in *Azospirillum*. The respiratory rates of *A. brasilense*, grown diazotrophically on a variety of C-substrates, were found to be similar to those of rhizobia, but only about 10 to 20% of those of *Azotobacter* (Okon *et al.* 1976b). The effect of O_2 concentration on nitrogenase activity of *Azospirillum* spp. yields typical bell-shaped curves (Day and Döbereiner 1976; Magalhães *et al.* 1983; Okon *et al.* 1976a; Pedrosa *et al.* 1981, 1982; Volpon *et al.* 1981), with variable optimum p_{O2}. Comparative analysis of these optimum p_{O2} for nitrogenase activity must be done with care, since they vary with bacterial density, respiratory activity and age, medium composition, temperature, size and geometry of flask, cell suspension/flask volume ratio, pressure, mode, and shaking rates, all factors which affect O_2 diffusion rates and consequently the actual dissolved O_2 concentrations. Optimum dissolved O_2 concentrations for diazotrophic growth and nitrogenase activity of *A. brasilense* determined with O_2 electrodes were reported in the range of 0.60 to 8 μM (0.0005 to 0.007 atm at 30°C) (Bergersen and Turner 1980; Nelson and Knowles 1978; Okon *et al.* 1977b). These O_2 values are sufficient to saturate the high affinity terminal oxidases (see below). Nevertheless, it is clear that an equilibrium between O_2 diffusion and respiration rates is essential to maintain optimal nitrogenase activities at these low O_2 values. Higher O_2 concentrations (>11 μM O_2) are not tolerated by the N_2-fixing systems of *A. brasilense, A. lipoferum*, and *A. amazonense* which behave similarly, although *A. amazonense* appears to be more sensitive to O_2 (Magalhães *et al.* 1983). *Azospirillum brasilense* grown diazotrophically at various dissolved O_2 concentrations had similar p_{O2} optima for nitrogenase activity demonstrating no adaptation to higher O_2 concentrations (Nelson and Knowles 1978), contrasting with those reported for *Azotobacter chroococcum* (see Postgate 1982).

Azospirillum spp. grown diazotrophically under air in a semisolid medium actively move to depths where the O_2 concentration gradient is most favorable for optimal N_2 fixation (Day and Döbereiner 1976). The aerotactic response of *A. brasilense* has been quantified (Barak *et al.* 1982; Okon *et al.* 1980) and found to depend on dissolved O_2. No aerotaxis occurred anaerobically and migration toward or away from O_2 depended on O_2 concentrations, culture age, incubation time, and respiration inhibitors. Aerotaxis was expressed both in the absence and the presence of combined nitrogen (see Section 2.1).

Catalase activity is present in most strains of *Azospirillum* spp. (Magalhães *et al.* 1983; Tarrand *et al.* 1978). Catalase activity of N_2- and NH_4^+-grown *A. brasilense* and of NH_4^+-grown *A. lipoferum* decreased with growth or increased cell densities (P. S. S. S. Diniz and F. O. Pedrosa, unpublished data). Similar results were reported for *A. brasilense* (Clara and Knowles 1984). This fall in catalase activity could not be reversed by increased dissolved oxygen tensions (Clara and Knowles 1984). However, catalase activity decreased, while superoxide dismutase (SOD) activity increased, with increasing dissolved oxygen concentrations in *A. brasilense* grown continuously in the presence of NH_4^+ (Nur *et al.* 1982b). The N source was found to affect the levels of catalase and peroxidase, but only slightly those of superoxide dismutase. Highest levels of catalase (52 to 80 units/mg protein) were found for *A. brasilense* grown in the presence of NH_4Cl and sparged with 0.0075 atm O_2 or in the presence of NO_3^- anaerobically, and lower levels (6 to 10 units/mg protein) were found in the presence of N_2 and 0.0075 atm O_2 or N_2O anaerobically (Clara and Knowles 1984). Superoxide dismutase activities were in the range of 6.2 to 9.1 units/mg protein with the four N sources utilized, and peroxidase activities were very low (0.03 to 0.08 units/100 mg protein). Further studies showed that *A. brasilense* peroxidase is particulate and that *A. brasilense* Sp7 grown on N_2, NO_3^-, or NH_4^+ had only one superoxide dismutase that was insensitive to CN^-, but was sensitive to H_2O_2 inhibition, thus indicating an Fe enzyme (Fe-SOD) (Clara and Knowles 1984). Moore *et al.* (1984), however, found that *A. brasilense* had one Mn-SOD and one Fe-SOD isozyme, while *A. lipoferum* had two Fe-SOD isozymes.

Although the presence of these enzymes should protect *Azospirillum* spp. against overall oxygen toxicity, a specific role of these enzymes in protecting nitrogenase of *Azospirillum* has not been demonstrated. A catalase overproducing (500- to 1000-fold) mutant of *A. brasilense* Sp7 showed no better resistance to inactivation and repression of nitrogenase by oxygen than the wild type (Higuti and Pedrosa 1985).

Carotenoid composition, synthesis, and function were studied in several strains of *A. brasilense* (Nur *et al.* 1981, 1982b). Carotenoids were present in strains Cd and Cd-1 and absent in strains Sp7, Sp81, and

Sp51e. Strains Cd and Cd-1 showed four and one carotenoid types, respectively. Synthesis of carotenoids was highest with 5 m*M* NH_4Cl and was not observed above 10 m*M* NH_4Cl. It did not occur under microaerobic conditions and was highest at high dissolved O_2 concentrations (Nur *et al.* 1981, 1982b). Diphenylamine, an inhibitor of carotenoid synthesis, decreased aerobic nitrogenase activity in strain Cd. These results led the authors to suggest that carotenoids may play a role in the protection of nitrogenase against O_2 (Nur *et al.* 1981). To check this hypothesis, carotenoid-negative mutants of *A. brasilense* strain Cd and carotenoid-overproducing mutants of strain Sp7 were isolated and compared (Hartmann *et al.* 1983a). A carotenoid-overproducing mutant strain C-1 had a carotenoid pattern similar to that described for strain Cd, and three of these mutants derepressed nitrogenase under air, while strain Sp7 failed to do so. In addition, strain C-1 showed slightly higher p_{O2} optima for nitrogenase activity than Sp7. This report apparently supports the suggestion that carotenoids may exert some protection against O_2. However, in none of these reports was the actual dissolved O_2 concentration for nitrogenase activity and derepression determined. Furthermore, carotenoid synthesis occurred only with high dissolved O_2 concentrations, conditions under which nitrogenase is not expected to be synthesized or active. Cell aggregation under aerobic conditions provides a microaerophilic environment which may be more important in protecting the N_2-fixing system from O_2 damage than carotenoids. It is worth mentioning that cell aggregation in *A. brasilense* strain Cd under aerobic conditions parallels carotenoid synthesis (Nur *et al.* 1981).

The nitrogenase enzyme Similar to other diazotrophs, the nitrogenase complex (Burgess 1984, 1985; Eady and Postgate 1974; Emerich and Burris 1978; Mortenson and Thorneley 1979; Postgate 1982) of *Azospirillum* spp. requires two Fe proteins for activity, an FeMo protein, and an Fe protein (Ludden *et al.* 1978), and nitrogenase catalyzes the reaction

$$N_2 + 8H^+ + 8e^- + 16\ ATP \rightarrow 2NH_3 + 16\ ADP + 16PO_4^{3-} + H_2$$

The Fe protein from *A. brasilense* and *A. lipoferum* is usually obtained in an inactive form that requires an activating enzyme for activity (Hartmann *et al.* 1985; Ludden *et al.* 1978; Okon *et al.* 1977b; Pedrosa and Yates 1984), similar to that described for *Rhodospirillum rubrum* (Ludden and Burris 1976). The Fe protein of *A. amazonense* does not require activation (Song *et al.* 1985) (see below). The nitrogenase proteins have been partially purified from *A. brasilense* (Ludden *et al.* 1978) and highly purified from *A. amazonense* (Song *et al.* 1985). The highly purified Fe and MoFe proteins of *A. amazonense* nitrogenase have been characterized (Song *et al.* 1985). Maximal activities of the Fe and MoFe proteins were 1,800 and 2,400 nmol C_2H_4/min/mg protein. The optimal combination of the Fe and MoFe protein for maximum nitrogenase activity of the Fe protein was at a molar ratio of 1 Fe to 2 MoFe protein and of the MoFe protein was at a molar ratio of 1 MoFe protein to 120 Fe proteins. The MoFe protein (dinitrogenase) has a molecular weight of 210,000, and the Fe protein (dinitrogenase reductase) a molecular weight of 67,000. The MoFe protein is tetramer ($\alpha_2\beta_2$) with subunits of 50,000 and 55,000 molecular weight, whereas the Fe protein is a dimer of subunit molecular weights of 31,000 and 35,000. The isoelectric points of the MoFe and Fe protein are 5.2 and 4.6, respectively. MoFe protein has 2 g-at. Mo, 24 g-at. Fe, and 28 g-at. acid-labile sulfide per mole. The Fe protein has 4 g-at. Fe and 6 g-at. acid-labile sulfide per mole (Song *et al.* 1985). Both proteins are extremely O_2 sensitive.

Azospirillum brasilense strain Sp7 nitrogenase component peptide molecular weights were determined following immunoprecipitation with antisera against *Klebsiella pneumoniae* Fe and MoFe protein (Nair *et al.* 1983). The MoFe protein polypeptides were estimated to have molecular weights of 60,000 and 64,000, whereas the Fe protein polypeptides have molecular weights of 33,000 and 36,000. Thus, the nitrogenase components of *A. amazonense* and *A. brasilense* have properties similar to those of other organisms (see Postgate 1982). The inactive Fe protein subunit of *A. brasilense* and *A. lipoferum*, however, showed an electrophoretic migration in sodium dodecyl sulfate (SDS)-polycrylamide gels similar to that of the modified subunit of *Rhodospirillum rubrum* (Hartmann *et al.* 1985, 1986). Although *A. amazonense in vivo* nitrogenase activity was switched off by NH_4^+ ions, no evidence was found for modification of its Fe protein (Hartmann *et al.* 1986). *Azospirillum amazonense* thus seems to lack the inactivating enzyme for Fe protein, being unable to promote ADP ribosylation of the Fe protein. Complementarity between *A. brasilense* and *A. amazonense* nitrogenase proteins and those of several other diazotrophs has been reported (Emerich and Burris 1978; Song *et al.* 1985).

Regulation of nitrogenase activity Nitrogenase activity of *Azospirillum* spp. cultures was found to be reversibly inactivated by NH_4^+ ions (Pedrosa and Yates 1983b), a process called "switch-on/off," similarly to what is observed in photosynthetic bacteria (see Zumft 1985, and references therein). L-Methionine sulfoximine, an inhibitor of glutamine synthetase (GS), abolished nitrogenase switch-off by NH_4^+ in *A. brasilense* and *A. lipoferum*, but not in *A. amazonense* (Hartmann *et al.* 1986). This suggests that glutamine or a product of GS activity triggers the switch-off mechanism. In contrast, nitrogenase activity of *A. brasilense ntr*C$^-$ mutants complemented by *K. pneumoniae nif*A^c was only slightly (25%) switched-off by NH_4^+ ions (Pedrosa and Yates 1983b) (see Section 2.4). At the molecular level the switch-off corresponds to an ADP ribosylation of the Fe protein of *A. brasilense* and *A. lipoferum*, but no such modification occurred in *A. amazonense* (Hartmann *et al.* 1986).

The Fe protein of *A. brasilense* and *A. lipoferum* is usually inactive when isolated (Hartmann *et al.* 1985; Ludden *et al.* 1978; Okon *et al.* 1977b; Pedrosa and Yates 1984), but can be activated by its own activating enzyme or by one from *Rhodospirillum rubrum* in a process requiring Mn^{2+} and ATP (Ludden *et al.* 1978). The Fe protein is thus inactivated by externally added NH_4^+ and also by the anoxygenic conditions prevailing during cell harvest (Hartmann *et al.* 1985).

NH_4^+ assimilation The enzymes involved in the primary assimilation of NH_4^+, glutamate dehydrogenase (GDH), glutamine synthetase (GS), and glutamate synthase (GOGAT), are present in *Azospirillum* spp. (Bani *et al.* 1980; Gauthier and Elmerich 1977; Okon *et al.* 1976b; Pedrosa and Yates 1983b, 1984). The GDH was found to be NADH dependent, while GOGAT was NADPH dependent. The GOGAT levels were higher (1.4 to 2.6 times) in N_2-grown cells than in NH_4^+-grown cells, while GDH levels were higher (1.5 times) in NH_4^+-grown cells. Glutamine synthetase activity, measured by the biosynthetic assay, however, was 8 times higher in N_2-grown cells than in NH_4^+-grown cells (Okon *et al.* 1976b). These results indicate that similarly to other diazotrophs, N_2-grown *Azospirillum* spp. assimilate NH_4^+ primarily via GOGAT and GS. In support of this hypothesis, GS and GOGAT deficient mutants of *A. brasilense* display a Nif^- phenotype (Bani *et al.* 1980; Gauthier and Elmerich 1977).

A repressible, energy-dependent transport system specific for NH_4^+ has been found in *A. brasilense* and *A. lipoferum* (Hartmann and Kleiner 1982). This system will transport methylammonium and NH_4^+ and is present in N_2-grown *A. brasilense* and *A. lipoferum*, but not in NH_4^+-grown cells. Aspartate and glutamate allow for the derepression of this transport system in *A. brasilense* but not in *A. lipoferum* (Hartmann and Kleiner 1982).

Glutamate synthase from *A. brasilense* has been purified and characterized (Ratti *et al.* 1985). The enzyme is an FeS flavoprotein with a stoichiometry of 1 flavin adeninedinucleotide:1 flavin mononucleotide:8Fe:8S per protomer molecular weight of 185,000. The protomer is composed of two dissimilar subunits with molecular weights of 50,000 and 135,000. Michaelis constant (K_m) values for NADPH, α-ketoglutarate, and L-glutamine were 6.25, 29, and 450 μM, respectively. The optimum pH for activity was 7.5.

Nitrate assimilation *Azospirillum brasilense, A. lipoferum*, and

A. amazonense grow aerobically using nitrate as the sole N source, and no nitrite accumulates in the culture medium (Bothe *et al.* 1981; Magalhães *et al.* 1978, 1983; Neyra and Van Berkum 1977; Neyra *et al.* 1977; Tarrand *et al.* 1978). *Azospirillum brasilense* Sp7 and *A. lipoferum* Sp59 are also capable of anaerobic growth using NO_3^- as the sole N source and terminal electron acceptor (Bothe *et al.* 1981; Nelson and Knowles 1978). *Azospirillum amazonense* is unable to grow under these conditions (Magalhães *et al.* 1983). Aerobic, microaerophilic, and anaerobic assimilation of nitrate involves assimilatory nitrate and nitrite reductases (Bothe *et al.* 1981; Nelson and Knowles 1978; Neyra and Van Berkum 1977; Neyra *et al.* 1977).

Denitrification Many strains of *A. brasilense* and *A. lipoferum* are able to denitrify (Eskew *et al.* 1977; Magalhães *et al.* 1978; Neyra and Van Berkum 1977; Neyra *et al.* 1977; Scott *et al.* 1979). Denitrification occurs anaerobically, under severe O_2 limitation, or microaerophilically (in semisolid medium or at $p_{O2} < 0.0075$ atm) (Eskew *et al.* 1977; Magalhães *et al.* 1978; Nelson and Knowles 1978; Neyra and Van Berkum 1977; Neyra *et al.* 1977; Scott *et al.* 1979). Under these conditions, nitrate is utilized as a terminal electron acceptor being reduced to nitrite. Many *A. lipoferum* strains reduce NO_3^- to N_2O with subsequent production of gas, whereas less than half of the *A. brasilense* strains tested produced gas from nitrite in semisolid medium or under anaerobiosis (Neyra *et al.* 1977). *Azospirillum brasilense* Cd and Sp7 were shown to denitrify to N_2, and many other strains accumulated N_2O under C_2H_2 (Eskew *et al.* 1977; Krieg and Döbereiner 1984). *Azospirillum amazonense* strains, however, are unable to carry out denitrification; no NO_3^--dependent N_2O production occurred under anaerobic conditions in the presence of NO_3^- nor did it occur in a semisolid medium under air (Magalhães *et al.* 1983).

Azospirillum brasilense strain Sp7, aside from growing anaerobically on nitrate, also grows anaerobically on nitrite as both a terminal electron acceptor and a N source (Zimmer *et al.* 1984) and on nitrous oxide (Stephan *et al.* 1984; Tibelius and Knowles 1984a) as an electron acceptor when supplied with NH_4^+ as an N source. Therefore, all steps of the anaerobic reduction of NO_3^- to N_2 are coupled to ATP production. Molar growth yield measurements indicate that the same amount of energy is conserved when *A. brasilense* Sp7 is grown on NO_2^- or O_2 as a terminal electron acceptor, while only one-third of the energy is conserved with NO_3^- (Danneberg *et al.* 1985). Nitrate dissimilation probably occurs via the pathway proposed by Matsubara (1970)

$$NO_3^- \underset{(1)}{\rightarrow} NO_2^- \underset{(2)}{\rightarrow} (NO) \underset{(3)}{\rightarrow} N_2O \underset{(4)}{\rightarrow} N_2$$

and involves four enzymes: (1) dissimilatory nitrate reductase, (2) dissimilatory nitrite reductase, (3) nitric oxide reductase, and (4) nitrous oxide reductase. All, with the exception of nitric oxide reductase, are present in *A. brasilense* and *A. lipoferum* (Bothe *et al.* 1981; Magalhães *et al.* 1978; Nelson and Knowles 1978; Neyra *et al.* 1977; Zimmer *et al.* 1984). Expression of these reductase activities in azospirilla follows complex regulation patterns involving O_2 concentrations, NO_3^- and NO_2^- concentrations, and pH (Nelson and Knowles 1978; Neyra and Van Berkum 1977; Neyra *et al.* 1977; Zimmer *et al.* 1984). The first step in denitrification occurs with NO_2^- accumulation in *A. brasilense* and *A. lipoferum* grown anaerobically or under O_2 limitation in a semisolid medium under air (Magalhães *et al.* 1978; Neyra *et al.* 1977). The nitrite formed by dissimilatory nitrate reductase is apparently excluded from, and is not toxic to, the bacteria (Neyra and Van Berkum 1977). Reduction of nitrite to gas (N_2O) occurs only after most of the nitrate has been exhausted (Magalhães *et al.* 1978; Nelson and Knowles 1978; Neyra and Van Berkum 1977). Nitrous oxide is reduced to N_2 by nitrous oxide reductase (Neyra and Van Berkum 1977; Neyra *et al.* 1977; Stephan *et al.* 1984). Nitric oxide production has never been observed from intact cells of *A. brasilense* (Zimmer *et al.* 1984). Nitric oxide is, however, produced in large amounts by cell-free preparations of NO_2^--grown *A. brasilense* (Danneberg *et al.* 1985). The authors considered this activity to be an artifact of solubilized dissimilatory nitrite reductase. This enzyme was found to be a cytochrome *cd*-containing reductase, although an additional Cu-containing reductase may also be present (Danneberg *et al.* 1985).

The levels of NO_3^-, NO_2^-, and N_2O reductases in anaerobically grown *A. brasilense* Sp7 were dependent on pH, nitrogen source, and oxygen concentration (Zimmer *et al.* 1984). The activity of NO_3^- reductase was high (12.3 to 17.8 μmol/min/mg protein) in the presence of NO_3^-, NO_2^-, and N_2O and in the range of pH 6.8 to 7.2, and it decreased at higher pH values (7.5 to 8.0). NO_2^- and N_2O reductase were depressed to very low values (0.2 to 1.5 μmol) in the presence of NO_3^- (10 m*M*) at pH values of 6.8 to 7.2, and they increased at pH 8 (3.6 to 4.6 μmol); high levels of both enzymes were present during growth on their respective substrates. Lower pH (7.0) or low concentrations of NO_2^- favored N_2O reductase and decreased NO_2^- reductase activities. Taken together the results suggest that denitrification in *A. brasilense* is controlled sequentially by the available nitrogen oxide intermediates: NO_3^- induces nitrate reductase and represses NO_2^- and N_2O reductases; NO_2^-, in turn, induces nitrite reductase and represses N_2O reductase synthesis, and N_2O induces high levels of N_2O reductase. This type of control has been proposed to occur in *A. brasilense* Sp7 by Nelson and Knowles (1978) from physiological studies.

NO_3^--dependent nitrogenase activity Several strains of *A. brasilense* and *A. lipoferum* showed NO_3^--dependent nitrogenase (C_2H_2 reduction) activity anaerobically or under severe O_2 limitation (Magalhães *et al.* 1978; Neyra and Van Berkum 1977; Neyra *et al.* 1977; Scott *et al.* 1979). Extensive studies of *A. brasilense* in continuous culture under various O_2 and NO_3^- concentrations failed to confirm nitrogenase activity dependent on NO_3^- respiration (Nelson and Knowles 1978). Transitory, nitrate-dependent C_2H_2 reduction by *A. brasilense* and *A. lipoferum*, however, has been confirmed in N_2-fixing, but not in NO_3^--grown cultures (Bothe *et al.* 1981). This activity was three to five times lower than that observed with O_2 respiration and stopped when the level of excreted NO_2^- reached 1 m*M*. It is doubtful whether this activity has any significance for azospirilla growth and anaerobic nitrogen fixation.

2.4 Genetics

Genetic studies in *Azospirillum* spp. began with the isolation of glutamine synthetase mutants (Gauthier and Elmerich 1977). The study organism was the newly-rediscovered diazotroph *Spirillum lipoferum* strain Sp7 (Döbereiner and Day 1976), now called *Azospirillum brasilense* (Tarrand *et al.* 1978). Progress in genetics has been relatively slow as compared to studies in physiology, ecology, microbiology, and agronomy of *Azospirillum* spp. Recently, however, the study of genetics of nitrogen fixation of these organisms has gained impetus with the isolation of *nif*-specific mutants (Jara *et al.* 1983; Pedrosa and Yates 1983b, 1984; Perroud *et al.* 1985; Singh and Klingmüller 1985) and the cloning of several genes (Fahsold 1985; Fogher *et al.* 1985a; Quiviger *et al.* 1982; Schwabe *et al.* 1985). Four species of *Azospirillum*, namely, *A. brasilense, A. lipoferum* (Tarrand *et al.* 1978), *A. amazonense* (Magalhães *et al.* 1983), and *A. halopraeferans* (Reinhold *et al.* 1986) are known. *Azospirillum brasilense* strains Sp7 and Sp6 are the organisms best studied genetically. Few data are available on the genetics of *A. lipoferum* and none on *A. amazonense* or *A. halopraeferans.*

Mutants of *Azospirillum* spp. The isolation of a collection of mutants is essential for the study of the overall genome of an organism and to determine the structural organization and regulation of specific metabolic pathways. So far, there is not a large number of *Azospirillum* mutants available, although auxotrophs, and antimetabolite- and antibiotic-resistant mutants can be easily isolated by the classic genetic procedures applied to *Escherichia coli* (Elmerich 1983).

Spontaneous, chemical- or radiation-induced mutants Mutants of *Azospirillum* spp., altered in several genetic traits, have been isolated.

These include auxotrophs and mutants resistant to antibiotics and antimetabolites deficient in ammonia assimilation, deficient in nitrate metabolism, resistant to hydrogen peroxide, deficient in carbohydrate metabolism, and deficient in nitrogen fixation.

Azospirillum spp. strains show variable low levels of natural resistance to antibiotics. *Azospirillum brasilense* Sp7 is resistant to nalidixic acid (4 μg/ml or 30 μg/disc), fusidic acid (10 μg/disc), trimethoprim (25 μg/disc), streptomycin (50 μg/ml or 10 μg/disc), kanamycin (4 μg/ml) (F. O. Pedrosa and M. G. Yates, unpublished data), carbenecillin, penicillin, and ampicillin bacitracin (Franche and Elmerich 1981, disc test, no concentrations given). This strain is sensitive, however, to tetracycline (2 μg/ml or 25 μg/disc), gentamicin (2 μg/disc), kanamycin (30 μg/disc), rifampicin (0.1 μg/ml), sulfatriad (200 μg/disc), chloramphenicol (25 μg/disc) (Franche and Elmerich 1981; F. O. Pedrosa and M. G. Yates, unpublished data), cefalotin, and cefaloridin (Franche and Elmerich 1981). Data on other strains are given elsewhere (Franche and Elmerich 1981; Magalhães *et al.* 1983; Polsinelli *et al.* 1980; Singh and Wenzel 1982). The great majority of *Azospirillum* spp. strains are resistant to high levels (400 μg/ml) of β-lactam antibiotics such as penicillin, ampicillin, carbenecillin (Franche and Elmerich 1981; Magalhães *et al.* 1983; Singh and Wenzel 1982). β-Lactamase activity was found in several *A. brasilense* and *A. lipoferum* strains (Franche and Elmerich 1981). The enzyme seems to be inducible since a growth lag around the antibiotic disc is observed (F. O. Pedrosa, unpublished data). The possibility of using this resistance as a taxonomic trait and in isolation procedures deserves consideration. Resistance to trimethoprim seems also to be common among *A. brasilense* and *A. lipoferum* strains (Singh and Wenzel 1982).

Azospirillum brasilense strain Sp7 has been mutagenized with *N*-methyl-*N*-nitro-*N*-nitrosoguanidine (NG), ethylmethane sulfonate (EMS), or ultraviolet (UV) radiation to obtain mutants resistant to rifampicin, streptomycin, and novobiocin (Franche *et al.* 1981; Jara *et al.* 1983). Treatment of exponentially growing cells of *A. brasilense* Sp7 with NG (20 μg/ml at 30°C for 30 min) increases the frequency of mutants resistant to kanamycin (20 μg/ml) 100-fold (F. O. Pedrosa and M. G. Yates, unpublished data), and in general, chemically (NG and EMS) induced mutants occur at a higher frequency (10^{-4}) (Jara *et al.* 1983; Pedrosa and Yates 1983b, 1984).

Attempts to raise the frequency of mutants resistant to 5-fluorotryptophan (140 μg/ml) and to rifampicin (5 μg/ml) by using UV and X rays, diepoxybutane, 4-nitroquinoline-*N*-oxide and 8-methoxypsoralen-UV, UV + acriflavine, or UV + 5-bromouracil failed, while treatment with EMS or NG was successful (Del Gallo *et al.* 1985). Based on these results, the authors concluded that *A. brasilense* lacks an SOS-like DNA repair mechanism.

Auxotrophic mutants of *A. brasilense* strains Sp6 (Bazzicalupo and Gallori 1983; Polsinelli *et al.* 1980), Sp7 (Franche *et al.* 1981; Gauthier and Elmerich 1977), and sp13t (Wood *et al.* 1982) have been obtained after NG, EMS, or UV mutagenesis followed by D-cycloserine enrichment. Single and multiple auxotrophs for both strains are available and have been used in studies on genome organization (see below).

Methylamine, an antimetabolite of ammonia, was found to be a weak growth inhibitor for *A. brasilense* strain Cd and *A. lipoferum* strain RG20a (Hartmann 1982). Mutants resistant to methylamine (50 m*M*) were isolated and found to be resistant in the presence of high (1 mg/ml) but not low (0.1 mg/ml) glutamate concentrations. The resistance observed in the presence of high glutamate concentrations was probably because metabolism of the mutant could establish elevated internal ammonium levels (Hartmann 1982). It remains to be determined whether these mutants have high derepressed levels of glutamate dehydrogenase.

Methionine sulfoximine (MSX) and methionine sulfone (MSF) are glutamate analogs which are inhibitors for both glutamine synthetase and glutamate synthase (Brenchley 1973). *Azospirillum bransilense* strains Sp7, Cd, and Sp35 and *A. lipoferum* SpRG6xx were shown to be very sensitive to MSX (0.1 μg/ml), while *A. lipoferum* strains RG20a and Br17 were resistant to MSX (up to 1 mg/ml) in the presence of NH_4^+ (10 m*M*); all strains were very sensitive to MSX in the absence of NH_4^+ (Hartmann 1982). Mutants resistant to MSX in the absence of ammonium ions were isolated after mutagenesis with NG (100 μg/ml at 30°C for 15 min) (Hartmann 1982).

Fluorotryptophan is an antimetabolite of tryptophan, a precursor of indole acetic acid synthesis in *Azospirillum* (Reynders and Vlassak 1979; Tien *et al.* 1979). Mutants resistant to 5-fluorotryptophan have been isolated following NG mutagenesis of several *A. lipoferum* and *A. brasilense* strains (Hartmann *et al.* 1983b) or as spontaneous Nif^+ revertants of Nif^- glutamate synthetase negative mutants of *A. brasilense* strain Sp6 (Barberio *et al.* 1982). Fluorotryptophan-resistant mutants of *A. brasilense* strain Cd (ATCC 29710) excrete high amounts of indole acetic acid, while *A. lipoferum* RG20a mutants fail to do so (Hartmann *et al.* 1983b). Similar mutants of *A. brasilense* strains Sp6 and Cd were also found to overproduce tryptophan (Barberio *et al.* 1982; Del Gallo *et al.* 1985).

Mutants resistant to sulfaguanidine, an antimetabolite of *p*-aminobenzoic acid and capable of excreting large amounts of *p*-aminobenzoic acid, have been isolated as revertants of NH_4^+ assimilatory mutants (Barberio *et al.* 1982).

Hydrogen peroxide is a toxic metabolite produced by aerobic metabolism. Aerobic cells produce catalase and peroxidase which, together with superoxide dismutase, have been suggested as participating in a mechanism for protecting aerobic cells against damage from oxygen and its reduction metabolites (Fridovich 1977; Halliwell 1981). A mutant of *A. brasilense* Sp7 capable of overproducing (500 to 1000 times) catalase has been isolated (Higuti and Pedrosa 1985).

Spontaneous mutants of *A. brasilense* strains Sp7, Sp13, and Sp34 and of *A. lipoferum* strain USA5a resistant to chlorate were isolated under oxygen-limited conditions (Magalhães *et al.* 1978). These mutants were found to lack nitrate reductase, nitrite reductase, or both activities. All nitrate reductase mutants were deficient in both anaerobic nitrite accumulation and aerobic NO_3^- utilization. This indicates the simultaneous loss of both assimilatory and dissimilatory nitrate reductase activities. However, only the dissimilatory nitrite reductase was lost in the selection with chlorate.

Azospirilla, similar to other diazotrophs (Postgate 1982), assimilate newly fixed ammonium via the glutamine synthetase–glutamate synthase pathway (Okon *et al.* 1976b), and mutants deficient in glutamine synthetase (Gauthier and Elmerich 1977) and in glutamate synthase (Bani *et al.* 1980) were found to display a nitrogen fixation negative (Nif^-) phenotype. Two classes of glutamine synthetase mutants were identified among glutamine auxotrophs of *A. brasilense* strain Sp7 (Gauthier and Elmerich 1977). The mutants of the first class (7028, 7037, and 7038) had low levels of glutamine synthetase activity (<25% of the wild type activity by biosynthetic assay or <6% by transferase assay in the presence of Mg^{2+}) and were able to derepress nitrogenase in the presence of high NH_4^+ concentrations (20 m*M*); these are nitrogen fixation constitutive mutants (Nif^c). The mutants of the second class (7029) were affected in both glutamine synthetase catalytic and "regulatory" properties, so that they could not derepress nitrogenase. Revertants recovered glutamine prototrophy (frequency 1×10^{-8}) and nitrogenase repressibility by NH_4^+. Recent work (Elmerich *et al.* 1985; Fogher *et al.* 1985c) showed that nitrogenase repression by NH_4^+ could be restored in mutant 7028 and 7029 by the introduction of plasmids containing the structural gene for glutamine synthetase (*gln*A) from *A. brasilense* or from *Klebsiella pneumoniae*. An active *gln*A gene product is an essential requirement to elicit nitrogenase repression by NH_4^+ in *K. pneumoniae* (see Dixon 1984). Bani *et al.* (1980) isolated and characterized *A. brasilense* Sp6 mutants lacking glutamate synthase activity. These assimilatory mutants also showed lower glutamine synthetase activities, were phenotypically Nif^- and were unable to utilize several N sources (histidine, alanine, xanthine, nitrate, and nitrite). Revertants able to fix nitrogen and to utilize the above-mentioned N sources (except nitrite) were isolated (Barberio *et al.* 1982).

Two mutants of *A. brasilense* strains Sp7, CW1, and CW2 were obtained after diethysulfate mutagenesis, and were shown to be severely deficient in both phosphoglycerate and pyruvate kinases; pleiotrophic mutation or multiple gene lesions have not been ruled out as responsible for the double lesions (Westby *et al.* 1983). No other mutants in carbon metabolism are known in *Azospirillum*.

At the time of writing, the following types of *nif* mutants of *A. brasilense* Sp7 were known: *nif*D$^-$ or *nif*K$^-$ (Jara *et al.* 1983; Pedrosa and Yates 1983b) and *nif*A$^-$ (Pedrosa and Yates 1983b, 1984). Recently, Tn5 insertional mutants *nif*K$^-$, *nif*D$^-$, and *nif*H$^-$ were described (Perroud *et al.* 1985). Two *ntr*C$^-$ (*gln*G$^-$) mutants of *A. brasilense* strains Sp7 (FP8 and 9) were characterized (Pedrosa and Yates 1983b, 1984). Both mutants are phenotypically Nif$^-$, unable to grow on nitrate as the sole N source and show decreased levels of glutamine synthetase (up to 30%). When complemented by plasmid pCK3 (*K. pneumoniae nif*A^c), both mutants recovered the Nif$^+$ phenotype and the ability to grow on nitrate. In addition, their nitrogenase was only partially susceptible (25 to 29%) to NH_4^+ switch-off *in vivo* (see Section 2.3).

Insertional mutagenesis Transposable elements (transposons) are discrete DNA segments flanked by repeated sequences and capable of moving from one replicon to another. Insertion of a transposon may cause gene inactivation or a mutation. Transposons are usually introduced into new hosts carried by plasmids unable to replicate in the new genetic background. These vectors are called "suicide" plasmids and are useful in promoting transposon mutagenesis. Attempts to obtain insertion mutagenesis in *A. brasilense* strain Cd by prophage Mu or by Tn5 carried by plasmids of the P incompatibility (IncP) group failed (Singh 1982). This failure was possibly due to the high stability of IncP plasmids in azospirilla (see below). Tn5::Mu insertional auxotrophs were, nevertheless, obtained when plasmid pJB4JI (IncP) was forced to segregate by plasmid pJB3JI (IncP) (Elmerich and Franche 1982).

More efficient systems to produce insertional mutants in *Azospirillum* spp. have been recently described (Perroud *et al.* 1985; Singh and Klingmüller 1985; Vanstockem *et al.* 1985).

Genetic elements The complete genome of an organism basically consists of chromosomes and plasmids. Additional genetic elements include insertion sequences, transposons, and lysogenic phages. For *Azospirillum* there are few studies on chromosome, plasmid, or bacteriophage size, content, function, and structural organization.

Chromosome There is no precise information on the size and number of chromosomes in *Azospirillum* species. Wood *et al.* (1982), using a modified Eckhardt agarose gel electrophoresis technique, detected various very high molecular weight DNA bands, which they called "minichromosomes." These minichromosome bands were found in several *A. brasilense* and *A. lipoferum* strains and varied in size from 42 to 1850 MDa (Megadaltons). For *A. brasilense* strain JM125, the simple sum of the molecular weights of the DNA bands totals 4,882 MDa, which is 1.8 times the molecular weight of the *E. coli* chromosome (Wood *et al.* 1982). Considering that the smaller DNA bands could be fragments from the larger ones and/or that the DNA bands might have overlapping regions, the chromosome of *Azospirillum* is possibly of a size similar to that of *E. coli.* The presence of minichromosomes in *Azospirillum* spp., instead of a single chromosome, has not been confirmed. These might be linear fragments of the chromosome, which entered the gels after 18 to 54 hours of electrophoresis. Although the presence of minichromosomes is questionable, the technique seems to give reproducible, strain-specific minichromosome patterns that could be useful for strain identification.

Indigenous plasmids Plasmids are extrachromosomal genetic elements capable of autonomous replication. Plasmids of various numbers and sizes have been found in *A. brasilense* and *A. lipoferum* (Franche *et al.* 1981; Heulin *et al.* 1982; Plazinski *et al.* 1983; Singh and Wenzel 1982; Wood *et al.* 1982) (Table 2.6). Methods of plasmid preparation using alkaline lysis give the most reproducible results. Each strain so far examined showed a unique set of plasmids in the range of 3.5 to 370 MDa. Clearly, the plasmid patterns are strain specific and cannot be used to distinguish *Azospirillum* species. Some strains have plasmids of similar sizes, although no comparative studies have been carried out to demonstrate their identities. Discrepancies in the number and size of plasmids reported for the same strain are probably due to plasmid instability, the technique used in their preparation, or to strain misnomers (Elmerich 1983). There are no reports on the plasmid content of *A. amazonense* and *A. halopraeferans.*

Several *Azospirillum* plasmids have been purified and cleaved by restriction enzymes (Franche and Elmerich 1981; Singh and Wenzel 1982). Plasmids of 27 and 46 MDa in *A. brasilense* (strain Kr77) and of 29 MDa (strain K67) have multiple *Eco*RI restriction sites, while plasmids of 3.5 and 13 MDa of *A. lipoferum* (strain SF50) and plasmid pAL1 (strains Br17 or RG20a) seem to have a single restriction site. Plasmid pAL1 also has a single restriction site for *Hin*dIII and multiple sites for *Hpa*I and *Hin*dIII (Singh and Wenzel 1982). Plasmids of low molecular weight and single restriction sites are potential candidates for cloning

Table 2.6 Indigenous plasmids of *Azospirillum* spp.

Strain	Plasmid molecular weight[a] (MDalton)	Reference[b]
A. brasilense		
Sp7	310, 310, 115, 90, 46	A
RO7	310, 310, 115, 90, 20, 4	A
K67	90, 30	A
KR77	115, 46, 27	A
Sp35	90	B
Cd	70	B
Sp7	78	B
Sp7	370, 350, 130, 100	C
Sp107	300, 280, 165, 150, 130, 40, 10	C
Sp245	130, 120, 100	C
UQ1774	130, 100	C
NA973	100	C
A. lipoferum		
Br17	150	A
DN64, 528	150, 115, 33, 29	A
SF50	150, 150, 115, 12, 3.5	A
SPN	150, 90, 8, 4.5	A
RG6xx	112, 49, 30, 8.4, 3.8	B
Br17, RG20a	36, 12, 5.4	B
Sp59	300, 280, 165, 150, 20	C
Sp242	165, 130, 120, 100, 30, 12	C
UQ1618	None	C
4B	> 270, 270, +3 plasmids > 120	D
4T	270, +2 plasmids > 120	D

[a]Plasmid patterns determined by agarose gel electrophoresis following alkaline lysis.
[b]References are as follows: A, Franche and Elmerich 1981, and Elmerich and Franche 1982; B, Singh and Wenzel 1982; C, Plazinski *et al.* 1983; D, Heulin *et al.* 1982.

vectors. Plasmid pAL1 was cloned into plasmid pACYC184 and pBR322 through their *Eco*RI and *Hin*dIII restriction sites, respectively. The hybrid plasmids (pAL1/pBR322 and pAL1/pACYC) were transformed into *E. coli* HB101 (Singh and Wenzel 1982). A 5.9-Kb plasmid of *A. brasilense* RO7 has also been cloned (Michiels *et al.* 1985).

Plasmids confer on the host cell phenotypic traits such as antibiotic and heavy metal resistance, antibiotic production, carbon source utilization, conjugational properties, bacteriocin and enterotoxin production, haemolysin production, induction of plant tumors, hydrogen sulfide production, host-controlled restriction and modification (see Hardy 1981), nitrogen fixation (Nuti *et al.* 1979), nodulation and host plant specificity (Johnston *et al.* 1978), H_2 utilization (Brewin *et al.* 1980), and others (see Hardy 1981; Foster 1983).

None of the indigenous plasmids of *Azospirillum* have so far been demonstrated to be conjugative, although some are large enough to harbor the complete set of *tra* genes (about 16 MDaltons) (Hardy 1981). No mobilizable plasmid has been found in *Azospirillum* spp., and there are no studies on plasmid copy number. Attempts to demonstrate amplification of the small plasmids of *A. brasilense* strains RO7, DN64, SF50, S28, and SpN with chloramphenicol were unsuccessful (Franche and Elmerich 1981). The bacteriocin-like activity found in several *A. brasilense* and *A. lipoferum* strains (Oliveira and Drozdowicz 1981) has not yet been reported as being plasmid borne.

Plasmid losses due to spontaneous instability (plasmids of 46 and 115 MDa of Sp7) or to treatment with acridine orange (Wood *et al.* 1982) were observed. *Azospirillum lipoferum* 4T, a spontaneous derivative of strain 4B lacking two plasmids (111 and 45MDa), lost its motility and the ability to produce the exopolysaccharide curdlan (Heulin *et al.* 1982). In a following report, Heulin *et al.* (1983) demonstrated high frequency of motility loss (0.9×10^{-3} per bacterium generation) and recovery (1.8×10^{-3}) for strains 4B and 4T, but made no mention of changes in the plasmid patterns. Similar changes in motility were also found in *A. lipoferum* strains Br10 and B7C, while *A. brasilense* strains Sp7, RO7, 51e, and A95 failed to show changes under similar conditions. It is tempting to suggest that changes in motility were associated with exopolysaccharide production, the polysaccharide being a specific requirement for motility. Inactivation of chromosomal genes involved in exopolysaccharide production by plasmid integration could explain the observed changes in motility. Alternatively, polar effects caused by plasmid integration could explain the loss of both traits, as pointed out by Heulin *et al.* 1983. An interesting and possibly related phenomenon is the reversible change from rough to smooth or from red to white (in the presence of Congo red) colonies, which has been observed in *A. brasilense* Sp245 and *A. lipoferum* Sp242 (J. Döbereiner, unpublished results). Curing of a large plasmid, (minichromosome pAZ4, 45 MDa, *A. brasilense* 125A2) with acridine orange led to increased sensitivity to Cd^{2+} ions and kanamycin, and to the loss of the ability to grow in ethanol or butanol (Wood *et al.* 1982).

Plasmid pAL1 from strain RG20a cloned into plasmid pBR322 or pACYC184 failed to confer *E. coli* HB101 with antibiotic resistance (carbenecillin, spectinomycin, trimethoprim, and tyrosidine), heavy metal resistance (Hg^{2+}, CO^{2+}, Zn^{2+}), or the ability to utilize various carbon sources (β-hydroxybutyrate, poly-β-hydroxybutyrate, xylose, or α-ketoglutarate) (Singh and Wenzel 1982). Plasmid pAL1 either may not carry genes for these traits or does not express them in *E. coli*.

Attempts to demonstrate the presence of nitrogen fixation (*nif*) genes in indigenous plasmids of *A. brasilense* and *A. lipoferum* through hybridization with *nif* genes from *Klebsiella pneumoniae* (see Elmerich 1983) or from *Rhizobium meliloti* (Plazinski *et al.* 1983) have failed. Planzinski *et al.* (1983) presented strong evidence that the *nif*HDK genes of *Azospirillum* are chromosomal. There are, however, reports indicating plasmid localization for *nif* genes in certain strains of *Azospirillum* (Uozumi *et al.* 1982; Zöphel *et al.* 1980). In spite of this evidence, no specific phenotype has been conclusively found to be plasmid borne in *Azospirillum* spp.

Bacteriophages Lysogenicity in *Azospirillum* spp. seems to be common. Mitomycin C (0.5 μg/ml) induced lysis in *A. brasilense* strains Sp7, Sp35, Sp13, and RO7 and in *A. lipoferum* strains SP 59b, BC3, USA5a, DN64, S28, SF50, and SpN, but failed to induce *A. brasilense* strains K67 and KR77 and *A. lipoferum* strains Br17 and RG20a (Elmerich *et al.* 1982; Franche and Elmerich 1981). Lysed strains are either lysogenic or carry information for bacteriocin production (Franche and Elmerich 1981). However, lysates from all strains were unable to form plaques on all other strains. Phagelike particles were partially purified from mitomycin C-treated *A. brasilense* Sp7 and shown to be noninfective and devoid of DNA (Franche and Elmerich 1981). *Azospirillum brasilense* Sp7 has been shown to lyse spontaneously after 2 to 4 weeks of daily subculturing in complex media. *Azospirillum brasilense* strains Cd and Sp35 and *A. lipoferum* strains Sp59b, RG20a, and Br17, all failed to lyse under these conditions. Infective phage particles (phage Ab-1) were purified from *A. brasilense* Sp7 lysates and were characterized (Germida 1984).

A temperate bacteriophage (Al-1) able to form turbid plaques on *A. lipoferum* strain Br17 was isolated from soil samples in Brazil (C. Rosemberg and J. Döbereiner, unpublished data), purified, and characterized (Elmerich 1983). The phage, similar in size and morphology to coliphage λ, has an icosahedral head 56 nm long and a tail 250 nm long with six spikes. Its genome consists of a linear double-stranded DNA molecule of 36 kilobase pairs showing cohesive ends, and in the lysogenic state it is maintained as a closed circular plasmid. No transduction properties have been demonstrated. The host range of phage Al-1 remains limited to *A. lipoferum* strains Br17, BC3, and RG59a and to *A. brasilense* strain SP35. *Azospirillum lipoferum* strains RG20a, USA5a, DN645, S29, SF50, and SpN and *A. brasilense* strains Sp7, Spl3, Cd, RO7, K67, and KR77 are insensitive to phage Al-1. Three naturally occurring mutants of *A. lipoferum*, two of strain Br12 (Br1702 and Br1710) and one of BC3 (BC31), are insensitive to Al-1. Spontaneous mutants of phage Al-1 forming clear plaques occurred at frequencies of about 10^{-4}. Further studies are needed to determine the extent of phage Al-1 host range and its potential as a tool in *Azospirillum* genetics.

Phage Ab-1 particles produced by *A. brasilense* Sp7 were shown to be infective to strains Sp7 and Cd, forming small clear plaques (0.5 to 1 mm), which become turbid with time. Strains Sp35, Sp59a, RG20a, and Br17 are resistant to phage Ab-1 (Germida 1984).

Genetic transfer Gene transfer systems are essential in order to establish the genetic make-up of an organism. These systems—bacterial conjugation, transformation, and transduction—allow the determination of the structure and regulatory organization of the genome.

Conjugation Conjugation was first demonstrated in *Azospirillum brasilense* strain Sp6 by Polsinelli *et al.* (1980). It is presently the most efficient gene transfer system for *Azospirillum* spp. Polsinelli *et al.* (1980) showed that plasmid pRD1, which carries the nitrogen fixation genes of *Klebsiella pneumoniae* (Dixon *et al.* 1976), could be transferred from *E. coli* strain JC5466 to *A. brasilense* strain Sp6 at a frequency of 1×10^{-7} per recipient. Another plasmid, R68.45, was transferred from *Pseudomonas aeruginosa* strain PAO25 to several auxotrophic mutants of *A. brasilense* Sp7 at a frequency of about 10^{-7} per donor (Franche *et al.* 1981). This plasmid was stably maintained in strain Sp7 for at least 50 generations in the absence of selective pressure and was transferred from *A. brasilense* Sp7 to Sp7 derivatives and to *A. lipoferum* Br17 derivatives at frequencies of about 5×10^{-2} per donor, respectively. Plasmid R68.45 derivatives pGE10, pGE50, and pGE51 (Espin *et al.* 1981) have also been transferred to *A. brasilense* Sp7 and its mutants at frequencies of 1×10^{-5} per recipient or 1×10^{-7} per donor (Pedrosa and Yates 1984). Plasmids R68.45 and RP4 (Bazzicalupo and Gallori 1983; Franche *et al.* 1981; Mishra and Roy 1981; Singh 1982) were shown to display chromosome mobilization ability (see below). Other plasmids of the P incompatibility group such as pCK3 (IncP, Tc, Km, and *K. pneumoniae nif*A^c) (Pedrosa and Yates 1983b, 1984), and suicide plasmids pJB4J1 (IncP, Gent, Mu, and Tn5), pJB3Ji (IncP, Tc, and Cb) (Elmerich and Franche 1982), and RP4::Mucts61 (Singh 1982) have also been introduced in *Azospirillum* by conjugation. A general observation is that plasmids of the P incompatibility group are stably maintained in *A. brasilense* and *A. lipoferum*, while plasmids of the Q and W groups are not (Michiels *et al.* 1985; Pedrosa and Yates 1984). A summary of the data on plasmid transfer to *Azospirillum* is presented in Table 2.7 (see pages 48–49).

Transformation Genetic transformation of *A. brasilense* strain Sp7 to antibiotic resistance (Mishra *et al.* 1979) and of *A. brasilense* strain 13st auxotrophic mutants to prototrophy (Wood *et al.* 1982) has been claimed. Total DNA (Mishra *et al.* 1979) or purified minichromosomes

(Wood *et al.* 1982) were used to transform mutants of the same strain. Maximum competence in *A. brasilense* occurred after a 6-hour growth period (10^6 cells/ml) in nutrient broth, and transformants for streptomycin, novobiocin, and kanamycin resistance appeared at frequencies of 1 to 3.4×10^{-6}. The appearance of transformants was proportional to the DNA concentration used and was eliminated by treatment with deoxyribonuclease (Mishra *et al.* 1979). The frequency of *A. brasilense* Sp7 transformation is very low relative to that of Mo-starved *Azotobacter vinelandii* (10^{-2} to 10^{-3}) (Page 1985) and close to the spontaneous mutation rate. Wood *et al.* (1982) used a Mn^{2+} low pH procedure to increase competence in *A. brasilense* strain 13st. Using total DNA or purified minichromosome bands, the authors claim to have transformed auxotrophic mutants back to prototrophy. Critical analysis of these results was not possible since no quantitative data were presented. Plasmid RP4 transformed *A. brasilense* Sp7 to simultaneous resistance to several antibiotics at frequencies of 2.5×10^{-6} (Mishra and Roy 1981). The uptake and stability of ^{32}P-labeled plasmid pUW610 (a hybrid between pACYC177 and a *Hin*dII fragment of pBR325 conferring chloramphenicol and kanamycin resistance) by *A. lipoferum* strain A23 have been studied (Schwabe and Weber 1983). Less than 1.4% of the initial plasmid DNA was adsorbed by the bacteria in a form resistant to DNase, but nothing was stated concerning the frequency of transformation of *A. brasilense* strain A23 to chloramphenicol or kanamycin resistance carried by plasmid pUW610. A mutant of this strain having 25% of the wild type nuclease activity showed no improvement in transformation competence.

In spite of these reports, several researchers have found it difficult to transform *Azospirillum* spp. Thus, the study of transformation in *Azospirillum* is still an open field, and the basic problem is still to improve cell competence.

Genetic mapping Attempts to determine the organization of chromosomal genes of *A. brasilense* strains Sp6 (Bazzicalupo and Gallori 1983) and Sp7 (Franche *et al.* 1981) have been carried out using the self-transmissible chromosome mobilizing plasmid R68.45 (Haas and Holloway 1978). Plasmid R68.45 transfer between *A. brasilense* strains occurs at frequencies of about 10^{-2} transconjugants per recipient, while selected markers (genes for a specific trait) were transferred at frequencies of 10^{-6} per recipient, independent of the marker selected (Franche *et al.* 1981). Plasmid R68.45 appears to promote nonpolar, bidirectional chromosomal gene transfer in *A. brasilense* strain Sp7, suggesting the presence of multiple sites of plasmid insertion in the chromosome of this strain (Franche *et al.* 1981), which is a situation analogous to that described for *Pseudomonas aeruginosa* (Haas and

Table 2.7 Conjugative plasmids transferred to *Azospirillum* spp. by conjugation

Plasmid	Bacterial donor	Bacterial recipient	Transfer frequency per recipient	Stability	Reference[a]
pRD1	*E. coli* JC5466	*A. brasilense* Sp6	10^{-7}		A
	E. coli JC5466	*A. brasilense* SPF66	10^{-7}		A
	A. brasilense SPF661	*E. coli* JC5466	10^{-5}		A
	A. brasilense SPF661	*A. brasilense* SPF94	10^{-1}	+(20)[b]	A
RP4 (IncP, tra, Km, Tc, Cb)	*E. coli* J5–3	*A. brasilense* Sp7	10^{-5}	+	B
	A. brasilense Sp7 (=JS1)	*E. coli* K12	3.7×10^{-5}		B
	A. brasilense Sp7 (=JS1)	*A. brasilense* Sp7 (S–200)	1.2×10^{-5}		B
	E. coli HB101	*A. brasilense* Cd	3×10^{-6}	+	C
	A. brasilense Cd	*E. coli*	10^{-4} to 10^{-6}		C
RP4::Mu (IncP, tra, Km, Tc, Cb ::Mu cts)	*E. coli* 921	*A. brasilense* Cd	3×10^{-6}	+	C
	A. brasilense Cd	*E. coli*	10^{-4} to 10^{-6}		C

R68.45 (IncP, tra, Km, Tc, Cb)	*Pseudomonas aeruginosa* PAO25	*A. brasilense* Sp7	10^{-7}/donor	+	D
	P. aeruginosa PAO25	*A. brasilense* Sp6	—[c]		E
	A. brasilense (Sp7 mutants)	*A. brasilense* (Sp7 mutants)	10^{-1} to 10^{-2}	+	D
	A. brasilense (Sp7 mutants)	*A. lipoferum* Br17	10^{-2}		D
	A. brasilense (Sp6 mutants)	*A. brasilense* (Sp6 mutants)	—	+	E
R68.45	*E. coli* 1230	*A. brasilense* (Sp7 mutants)	10^{-5}	+	F
pJB4JI (IncP, Gent, Mu, Tn5, tra)	*E. coli*	*A. brasilense* Sp7	—	+	G
	E. coli	*A. brasilense* Cd	5×10^{-6}	+	C
pJB3JI (IncP, Tc, Cb, tra)	*E. coli* RI	*A. brasilense* Sp7 (pJB4J1)	—	+	G

[a]References are as follows: A, Polsinelli *et al.* 1980; B, Mishra and Roy 1981; C, Singh 1982; D, Franche *et al.* 1981; E, Bazzicalupo and Gallori 1983; F, Pedrosa and Yates 1984; G, Elmerich and Franche 1982.

[b]Number of generations in the absence of selective pressure.

[c]Dashes indicate frequency not reported.

Holloway 1978). Similar results were obtained by Bazzicalupo and Gallori (1983) using *A. brasilense* strains Sp7 and Sp6 (a strain not related to Sp7).

Although map distances and tentative order for several markers have been obtained (Bazzicalupo and Gallori 1983; Franche *et al.* 1981), the data so far obtained are not sufficient to draw a genetic map for the chromosome of *A. brasilense*. Difficulties in constructing the genetic map of this organism are probably due to (a) low frequency of marker transfer, (b) nonpolar transfer of markers giving nonsequential gene transfer, (c) multiple site of plasmid R68.45 attachment giving bidirectional multiple origin of transfer initiation, and (d) formation of multiple R68.45 prime episomes. The use of other plasmids with enhanced polar chromosomal mobilization activity showing few sites of attachment should help to determine the gene organization of the chromosome of *A. brasilense*. Alternatively, the isolation of specific R68.45 primes, carrying pieces of *Azospirillum* chromosomes, could be valuable in this respect.

Transformation and transduction have not been applied to determine the genetic map of *Azospirillum* spp. There are no reports on genetic mapping of *A. lipoferum, A. amazonense*, or *A. halopraeferans*.

***Nif* genetics of *Azospirillum* spp.** To understand the recent development in the genetics of nitrogen fixation (*nif*) in *Azospirillum* spp., the organization, function, and regulation of the nitrogen fixation genes of *Klebsiella pneumoniae* is considered here. For detailed accounts of the development of *nif* genetics of *K. pneumoniae*, see Cannon *et al.* (1985), Dixon (1984), Postgate (1982), Robson *et al.* (1983).

Klebsiella pneumoniae is the best studied diazotroph in terms of *nif* genetics and is taken as the basic model. A landmark in the development of *nif* genetics of this organism was the *in vivo* construction of plasmids containing the whole set of *nif* genes (Dixon *et al.* 1976; Postgate 1982). The techniques of classic genetics and recombinant DNA brought *nif* genetics of *K. pneumoniae* to its present stage of development. Expression of nitrogen fixation in *K. pneumoniae* strain M5a1 requires 17 *nif* specific genes organized into 7 or 8 operons. The *nif* operons constitute a contiguous cluster occupying a DNA segment of about 23-Kb pairs in length located in the chromosome between the *shi*A (shikimate permease) gene and the *his*D (L-histidinol:NAD^+ oxidoreductase) gene (Figure 2.7). The sizes and functions of the *nif* gene products of *K. pneumoniae* are shown in Table 2.8, and described as follows. The *nif*B gene product is involved in the synthesis or insertion of the FeMo cofactor (FeMoco). Present evidence indicates that the *nif*Q gene product may be involved in molybdenum transport. The products of *nif*L and *nif*A have regulatory functions at the transcriptional level of *nif* expres-

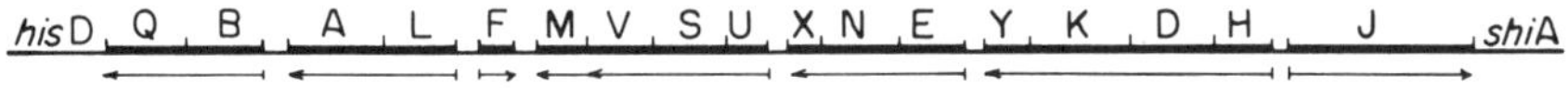

Figure 2.7 Map of the nitrogen-fixation (*nif*) operons of *Klebsiella pneumoniae* M5a1. Bars indicate operon transcripts. Arrows indicate direction of transcription. (Robson *et al.* 1983; Cannon *et al.* 1985).

sion. Under limiting NH_4^+ or O_2 levels, the product of *nif*A activates all *nif* promoters. However, when NH_4^+ or O_2 levels are high, the product of *nif*L blocks activation by the *nif*A gene product. *Nif*A and *nif*L gene products exert a *nif* specific transcriptional regulation. The *nif*LA operon is under control of the general nitrogen source utilization system (*ntr* or *gln*) (Merrick 1983). The product of *nif*F is a flavodoxin. In the *nif*M operon the product is possibly involved in posttranslational modification of the Fe protein. The gene products of the *nif*USV operon are involved in posttranslational modification of the MoFe protein. *Nif*M can

Table 2.8 The *nif* gene products of *K. pneumoniae* strain M5a1: molecular weight and functions[a]

Operon	Gene	Molecular weight (KDaltons)[b]	Function
BQ	B	49	Synthesis or insertion of FeMoco
	Q	?	Molybdenum transport
LA	L	50	Repressor of *nif* transcription
	A	60	Activator of *nif* transcription
F	F	20	Flavodoxin–electron transfer to nitrogenase
M	M	28	Maturation of Fe protein
USV	U	25	Maturation of MoFe protein
	S	45	Maturation of MoFe protein
	V	42	Modifies substrate specificity of MoFe protein, synthesis of FeMoco
ENX	E	40	Synthesis or insertion of FeMoco
	N	50	Synthesis or insertion of FeMoco
	X	18	Unknown
HDKY	H	35	Subunit of Fe protein
	D	56	α-Subunit of MoFe protein
	K	60	β-Subunit of MoFe protein
	Y	24	Unknown
J	J	120	Pyruvate–flavodoxin oxidoreductase subunit

[a]Compiled from Robson *et al.* 1983, and Cannon *et al.* 1985.

[b]All gene products are polypeptides.

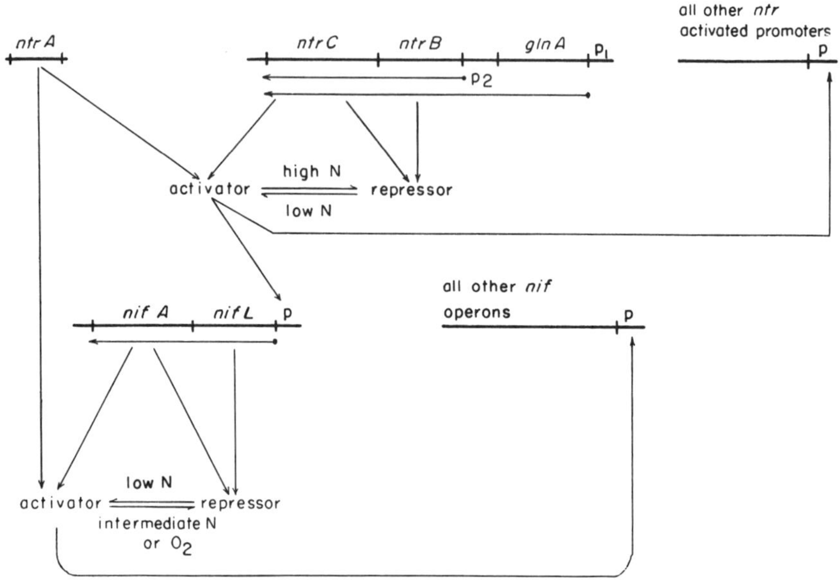

Figure 2.8 Merrick's model for *nif* regulation in *Klebsiella pneumoniae.* (Merrick 1983; Dixon *et al.* 1984)

be transcribed from its own promoter or from the *nif*U promoter. The *nif*EN gene products, as that of *nif*B, are needed for FeMoco synthesis or insertion into the MoFe protein. Derepressed extracts of mutants in these genes are complemented *in vitro* by purified FeMoco. The function of the *nif*X gene product is unknown. In the *nif*HDKY operon *nif*H codes for the Fe protein and *nif*K and D code for the α and β subunits of the MoFe protein of nitrogenase. The *nif*Y gene codes for a 19 to 24 KDa protein whose function is also unknown. The *nif*J gene product is the enzyme pyruvate-flavodoxin oxidoreductase, which reduces the flavodoxin product of *nif*F, using pyruvate as electron donor. The reduced flavodoxin is the electron donor to the Fe protein of nitrogenase.

Expression of nitrogen fixation in *K. pneumoniae* is under dual control: a *nif*-specific system involving the *nif*LA gene products and a nitrogen utilization system involving the *ntr(gln)* gene products. The model proposed by Merrick (1983) integrates both controls and has been largely confirmed (Figure 2.8). According to this model, under N limitation the product of *ntrC* (*glnG*) together with the product of *ntrA* (*gln*F., a σ factor) activates transcription of the *nif*LA operon. The product of *nif*A, together with that *ntr*A activates transcription of all other *nif* promoters. At intermediate levels of NH_4^+ the product of *nif*L blocks activation of *nif* transcription by the product of *nif*A. A possibly analogous situation occurs at high levels of fixed N when the product of *ntr*B inactivates the product of *ntr*C. *Klebsiella pneumoniae* mutants lacking active *ntr*A, *ntr*C, or *nif*A gene products are phenotypically deficient in nitrogen fixation and are termed regulatory mutants. Due to structural and functional similarities between *nif*A and *ntr*C gene products, *ntr*C$^-$ mutants can be complemented by *nif*A$^+$. Repression of nitrogenase synthesis by O_2 also appears to be mediated by the *nif*L gene product (Cannon *et al.* 1985; De Bruijn *et al.* 1984; Dixon 1984; Merrick 1983).

Localization of nif *genes* There is good evidence that the *nif* genes of *A. brasilense* and *A. lipoferum* are located in the chromosome (Planzinski *et al.* 1983) (see indigenous plasmids, Section 2.4). There are, however, no reports on which chromosomal genes the *nif* genes flank or whether they constitute a contiguous cluster, as in *K. pneumoniae.*

Although auxotrophic and antibiotic resistant mutants of *Azospirillum* are easy to isolate, *nif*-specific mutants are difficult. This difficulty relates to the nondistinctive character of *nif*$^-$ colonies on N-free agar medium, except for their smaller size. In addition, *Azospirillum* mutants phenotypically deficient in nitrogen fixation (Nif$^-$) may occur due to mutations affecting respiration, motility, carbon metabolism, NH_4^+ assimilation, N source utilization, etc., thus decreasing the frequency of *nif*-specific mutants in the mutagenized population. The first *nif*-specific mutants of *A. brasilense* were isolated only recently. Two apparently similar mutants in the nitrogenase structural genes *nif*D$^-$ and *nif*K$^-$ were characterized by biochemical (mutant FP6, Pedrosa and Yates 1983b; mutant 7571, Jara *et al.* 1983) and genetic complementations (Jara *et al.* 1983). Three regulatory mutants, one *nif* specific (*nif*A) and two of the general N utilization system (*ntr*C$^-$-like mutants), were also characterized (Pedrosa and Yates 1984). Insertional mutants of the *nif*HDK genes have also been isolated (Perroud *et al.* 1985).

Homology between Azospirillum *spp. DNA and cloned* nif *genes* DNA fragments from *Klebsiella pneumoniae* and *R. trifolii*, carrying the whole or part of the *nif*HDK operon, were used as hybridization probes to detect the presence of homologous genes in *Azospirillum* spp. Plasmid pSA30 (Riedel *et al.* 1979) carrying the *nif*HDK operon, and plasmid pW564 carrying *R. trifolii nif*HD genes (Plazinski *et al.* 1983) or the correspondent *nif* fragments. ^{32}P-labeled, successfully hybridized to total DNA from all *Azospirillum* strains tested (Elmerich and Franche 1982; Fahsold *et al.* 1985; Plazinski *et al.* 1983; Quiviger *et al.* 1982; Wenzel *et al.* 1983). The sizes of *Eco*RI and *Hin*dIII DNA fragments from various *Azospirillum* strains homologous to the *nif*HDK genes are presented in Table 2.9. The 24-Kb *Hin*dIII fragments from *A. brasilense*

Table 2.9 Sizes of *Eco*RI and *Hind*III DNA fragments from *Azospirillum* spp. homologous to the *nif*HDK operon

Strain	*Eco*RI (Kb)	*Hind*III (Kb)	Probe and reference[a]
Sp7	6.7	24	A
RO7	16, 2.35	18	A
K67	7.4	22.5	A
Br17	16, 1.8	15	A
S28	9.8	6.4	A
Sp7	18	—	B
Sp107	18	—	B
UQ774	18	—	B
Sp245	19	—	B
NA973	23	—	B
Sp59	15.5	—	B
Sp242	15.5	—	B
UQ1618	15.5	—	B
Cd	6.7	—	C
A23	9, 2.5	—	C
RG20a	2.5, 5.1, 6.1	—	D
Br17	2.5, 5.1, 6.1	—	D
RG6xx	5.0, 7.5	—	D
Sp7	6.7	—	D
Sp35	6.7, 9.4	—	D

[a]The ^{32}P-labeled hybridization probes and references are as follows: A, Purified *nif*HDK fragment of *K. pneumoniae* from pSA30, Quiviger *et al.* 1982, and Elmerich and Franche 1982; B, plasmid pW564, Plazinski *et al.* 1983; C, plasmid pAB1, Wenzel *et al.* 1983; D, purified AbRI fragment, Fahsold *et al.* 1985.

Sp7 DNA contains the 6.7-Kb *Eco*RI fragment homologous to *K. pneumoniae nif*HDK (Quiviger *et al.* 1982). Nair *et al.* (1983) studied the homology between plasmids which span the whole *K. pneumoniae nif* cluster and total DNA from *A. brasilense* strain Sp7 and *A. lipoferum* Br17. Weak homology was found between total DNA from both species and plasmids pPC936 and pMC71A which carry *K. pneumoniae nif*-FLABQ and *nif*A plus part of *nif*L genes. Homology was also detected between plasmid pPC880, which carries *K. pneumoniae nifJ* and *A. lipoferum* Br17 DNA; and no such homology was found with *A. brasilense* Sp7 DNA. Finally, no homology was found between plasmids pPC937 (*K. pneumoniae nif*XUSVM) or pGR113 (*K. pneumoniae nif*EN) and both *Azospirillum* species.

Nif genes of *Azospirillum* spp. have been cloned using the techniques of recombinant DNA. *Azospirillum brasilense* and *A. lipoferum* gene banks of total DNA have been constructed in phages λ*gt*7-*ara*6 (strain

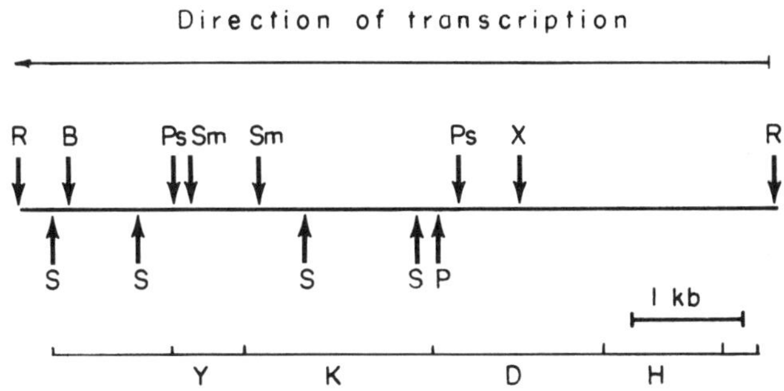

Figure 2.9 Physical map of the AbRI fragment. Restriction sites: B, *Bgl*II; P, *Pvu*II; Ps, *Pst*I; R, *Eco*RI; S, *Sal*I; Sm, *Sma*I; X, *Xho*I (Quiviger *et al.* 1982; Nair *et al.* 1983).

Sp7, Quiviger *et al.* 1982), λ47.1 (strains Cd and A23, Wenzel *et al.* 1983), in vectors, in plasmid pRK290 (strain Sp6, Fani *et al.* 1985), and in cosmids (strains RG20a and SpRG6xx, Fahsold *et al.* 1985; Sp7, Fogher *et al.* 1985a; and Sp35, Schwabe *et al.* 1985). The nitrogenase structural genes (*nif*HDK) have been cloned and analyzed. A recombinant phage, λ*gt*-Ab101, carrying the 6.7-Kb *Eco*RI fragment from *A. brasilense* Sp7, termed AbRI which is homologous to a *K. pneumoniae nif*HDK probe, was identified (Quiviger *et al.* 1982). Fragment AbRI was subcloned at the *Eco*RI site of plasmid pACYC184, forming plasmid pAB1. Restriction endonuclease analysis of AbRI showed that it contains sites for the enzymes *Bgl*II, *Pst*I, *Pvu*II, *Sal*I, *Sma*I, and *Xho*I, but no sites for *Kpn*I, *Xba*I, *Hin*dIII, *Hpa*I, and *Bam*H1 (Quiviger *et al.* 1982) (Figure 2.9). AbRI was also subcloned in the *Eco*RI site of the wide host range plasmid pRK290 forming plasmid pAB35, which complemented strain 7571, a *nif*D⁻ or *nif*K⁻ mutant of strain Sp7 (Jara *et al.* 1983). Plasmid pAB36 is a derivative of pAB35 with a 2.6-Kb *Pst*I fragment deleted from the AbRI fragment. The *Pst*I fragment deleted from plasmid pAB35 likely contains *nif*K and part of *nif*D gene of Sp7. Plasmid pAB36 possibly codes for *nif*H, although it did not complement several *A. brasilense* Sp7 *nif*H⁻::Tn5 mutants (Jara *et al.* 1983; Perroud *et al.* 1985).

Heteroduplex DNA analysis of AbRI revealed that the *nif*HDK genes of *A. brasilense* are transcribed in the same direction and constitute a single operon as in *K. pneumoniae* (Quiviger *et al.* 1982). These conclusions are also supported by transcriptional–translational analysis and genetic complementation of *A. brasilense* Sp7 *nif*HDK Tn5 insertion mutants (Perroud *et al.* 1985). The nitrogenase structural genes of *A. lipoferum* strains RG20a and RG6xx were also cloned, restriction nuclease analyzed, and found to be clustered and organized in the same orientation as in *A. brasilense* Sp7 (Fahsold *et al.* 1985).

Although homology has been detected between *K. pneumoniae nif*A and *nif*J and *Azospirillum* DNA (Nair *et al.* 1983), these genes have not yet been cloned. Recently, a DNA fragment from *A. brasilense* Sp7 contiguous to *nif*Y was cloned into plasmid pRK290 producing plasmid pAB131. This latter plasmid contains *nif*E and possibly other *nif* genes, such as *nif*N (Elmerich *et al.* 1985). Although initially, weak or no homology was detected between *K. pneumoniae nif*LABQ and *nif*-ENXUSVM genes and *Azospirillum* DNA (Nair *et al.* 1983), the cloning of *A. brasilense* Sp7 *nif*E seems to indicate that a much wider similarity in the organization of the *nif* clusters of these organisms exists.

The glutamine synthetase (*gln*A) gene of *A. brasilense* Sp7 was cloned into cosmid pVK100 and characterized (Fogher *et al.* 1985a,c). The derived plasmid pAB450 abolished the glutamine requirement of *A. brasilense* mutants 7028 and 7029, and restored their normal control of nitrogen fixation. The product of the *A. brasilense gln*A gene was found to be a 50-Kda polypeptide. Class II restriction endonuclease activity from *A. brasilense* Sp35, termed AbrI, was found to be a isoschizomer of XhoI, specific for the nucleotide sequence 5′C↓TCGAG. The gene coding for AbrI has been cloned (Schwabe *et al.* 1985). *Azospirillum amazonense* QR42 produces a similar enzyme (Schwabe *et al.* 1985). The plasmid pAF58 containing *A. brasilense* Sp6 *his* genes carried by a 9-Kb DNA fragment inserted into the *Eco*RI site of pRK290 was selected by its ability to complement *Azospirillum brasilense* and *Escherichia coli* His$^-$ auxotrophs (Fani *et al.* 1985). According to the authors, this *Azospirillum* DNA fragment should contain the genes coding for the enzymes imidazole glycerol phosphate dehydrase and histidinol phosphate phosphatase.

Homology between *Azospirillum* DNA and other genes relevant to nitrogen fixation were reported (Elmerich *et al.* 1985). Fogher *et al.* (1985b) found homology between genes determining host plant specificity (*hsn*) and nodulation (*nod*) of *Rhizobium meliloti* and DNA from *Azospirillum*. DNA homology was also found between the *fix*A gene of *R. japonicum* and *Azospirillum* DNA.

Studies on the regulation of *nif* expression have been carried out only in *A. brasilense*. The nitrogen-fixing system of *Azospirillum* spp., like those of other diazotrophs (see Postgate 1982), is not expressed in the presence of high levels of NH_4^+ or oxygen (Day and Döbereiner 1976; Döbereiner and Day 1976; Okon *et al.* 1976a). The molecular basis of this phenomenon in *K. pneumoniae* is well known and involves transcription regulation mediated by products of the *nif*LA operon and the *ntr(gln)* genes (Merrick 1983). Evidence from genetic complementation studies indicates that a similar regulation system operates in *A.*

brasilense strain Sp7 (Pedrosa and Yates 1983b, 1984). Three Nif^- mutants (FP8, FP9, and FP10) were found to be complemented by plasmid pCK3, which carries the *K. pneumoniae nifA* gene expressed constitutively. *Nif* expression in these transconjugants was constitutive in the presence of NH_4^+ and occurred at 30°C but not at 37°C due to the thermolability of the *K. pneumoniae nifA* gene product (Buchanan-Wollaston *et al.* 1981). Since *Azospirillum* fixes nitrogen optimally at 35 to 37°C (Day and Döbereiner 1976), one can conclude that the *A. brasilense nifA* gene product is thermostable in this temperature range.

Participation of the general nitrogen regulation (ntr) system in *nif* expression in *A. brasilense* was evident when two of the mutants (FP8 and FP9) were also complemented by *K. pneumoniae ntrC* gene (Pedrosa and Yates 1984) carried by plasmid pGE10 (Espin *et al.* 1981). These mutants are *ntr*C^- mutants which, as in *K. pneumoniae*, can be corrected by *nifA* (Buchanan-Wollaston *et al.* 1981; Merrick 1983; Ow and Ausubel 1983). The involvement of *nifA* in *nif* expression has been confirmed in another *A. brasilense* strain Sp245 (Funayama *et al.* 1985).

2.5 Ecology

Spirillum lipoferum was described by Beijerinck in 1925 and found to occur in most of the soil samples collected shortly after that in Germany and Austria (Schröder 1932). Surprisingly, the existence of this strain was lost and completely forgotten to such an extent that it was omitted from *Bergey's Manual of Determinative Bacteriology* (1974). The "nitrogen-fixing vibrio or spirillum" isolated by Becking (1963) from various African soils was not clearly identified as *S. lipoferum*, although he suggested a possible relationship. The introduction of semisolid, N-free media (Döbereiner and Day 1976) led to the discovery that this bacterium is one of the most widely distributed aerobic, N_2-fixing bacteria, especially in warm regions (Table 2.10). More than 60% of nearly one thousand grass roots and about 200 soil samples from tropical Africa and South America contained azospirilla in numbers ranging usually from 10^5 to 10^7/g of root or soil.

Azospirillum spp. have been isolated from rhizosphere soil and roots from all major cereals, including from:

1. Maize and sorghum in Brazil (Baldani 1984; Bülow and Döbereiner 1975), India (Subba Rao 1981), Egypt (Hegazi *et al.* 1979), and several temperate climatic regions (Vlassak and Reynders 1978), where it was found even in roots of already harvested sorghum (Wong and Stenberg 1979);

Table 2.10 Occurrence of *Azospirillum* in soils and roots[a]

Location	Latitude	Roots		Soil	
		No.	% positive	No.	% positive
Canada	50–60°N[b]	—	—	17	35
Europe	40–500°N	—	—	109	55
United States	35–50°N	62	11	11	64
New Zealand	40°S	—	—	3	33
Brazil	30°S	52	30	—	—
Brazil	0–23°S	926	62	144	59
Africa	6–15°N	45	58	47	89

[a]From Döbereiner *et al.* 1976, and Vlassak and Reynders 1978.
[b]Nine samples from Iceland were negative.

2. Wheat in Brazil (Baldani and Döbereiner 1980), India (Kavimandan *et al.* 1978), and Belgium (Vlassak and Reynders 1978);
3. Paddy rice in India (Kulasooriya *et al.* 1981; Nayak and Rao 1977), the Philippines (Watanabe *et al.* 1979), and France (Bally *et al.* 1983);
4. Dry rice in Brazil (Baldani 1984).

They have also been found in roots of aquatic weeds associated with rice (Nayak *et al.* 1979) and in the leaf sheath of deep-water rice (Kulasooriya *et al.* 1981). They are very common in the rhizosphere and roots of tropical forage grasses in Central and South America and South Africa (Tyler *et al.* 1979), and have been isolated from cold climate grasses in Finland (Haahtela *et al.* 1981) and Poland (Kulinska 1983). Associations with sweet potatoes (Hill *et al.* 1983), several tropical fruit trees (Subba Rao 1983), and Amazonian palm trees (Magalhães and Döbereiner 1984) were also reported. An organism later identified as *A. lipoferum* (Falk *et al.* 1986) was even isolated from lake water (Skerman *et al.* 1983). The relatively fewer reports from temperate regions and their less frequent occurrence (Döbereiner *et al.* 1976) might be explained by the more intensive agricultural practices and high fertilizer levels that predominate there. However, when samples from unfertilized plots were examined, azospirilla occurrence and numbers were still much lower in temperate than in tropical regions (J. Döbereiner, unpublished observations). Israeli soils do not contain native azospirilla possibly because of the pronounced dry season (Y. Okon, personal communication).

The availability of carbon substrates is usually thought to limit multiplication of N_2-fixing bacteria in soil in competition with the rest of

Table 2.11 **$^{15}N_2$ incorporation (μg/bottle) by *Azospirillum* spp. in sand amended with carbon substrates for 7 days[a]**

		A. lipoferum	
	A. brasilense Sp7	USA5A	101 (Australian isolate)
Malate	48	156	84
Xylan	4	59	17
Straw	11	32	31
Cellulose	0	0	0

[a]From Halsall *et al.* 1985.

the soil microflora, especially with organisms able to metabolize cellulose. Recently *A. lipoferum* and *A. brasilense* have been found to fix N_2 in mixed cultures with cellulolytic bacteria using cellulose as the sole carbon source (Halsall and Gibson 1985; Wong *et al.* 1980). Stable associations of *Azospirillum* with *Cellulomonas* could be established which fixed 15 mg N_2/g cellulose in 2 weeks.

Several Australian strains of *A. lipoferum* and *A. brasilense* were shown to utilize wheat straw and xylan as carbon sources for N_2 fixation in pure culture (Halsall *et al.* 1985) (Table 2.11). In addition, the production of pectolytic enzymes by these organisms (Tien *et al.* 1981; Umali-Garcia *et al.* 1980) indicates their wide range of metabolic diversity, as would be expected from organisms so widely distributed in the rhizosphere and in soils.

Aside from the availability of carbon substrates, various intrinsic characteristics of the bacteria and the remaining soil microflora affect growth and persistence of azospirilla within the microbial ecosystem. A large number (56) of actinomycetes, isolated from tropical soils, showed antagonistic properties against all or some *A. lipoferum* strains tested, while *A. brasilense* strains were less affected (Kulinska and Drozdowicz 1983). In this study, only 10% of the soil fungi population inhibited growth of azospirilla. The frequent occurrence of low-level streptomycin-resistant *Azospirillum* strains, especially in the rhizosphere and on root surfaces and interiors, where selection for low-level streptomycin resistance seems to occur for *Azospirillum* and other rhizosphere bacteria (Baldani and Döbereiner 1980; Brown 1961; Döbereiner and Baldani 1979), apparently confers *Azospirillum* with selective advantages (Table 2.12). Increased numbers of low-level streptomycin-resistant bacteria in the rhizosphere of maize, wheat, and beans were shown to occur, and the possibility exists that assimilation and translocation of this antibiotic by plant roots also occurs (Baldani *et al.* 1982). High numbers of actinomycetes were found on the root surface,

Table 2.12 Enrichment of low-level streptomycin-resistant bacteria in field-grown maize roots[a]

	Total no. bacteria per g dry wt × 10^7	% resistant to 20 μg/ml streptomycin
Soil A		
Rhizosphere soil	39 ± 27	1.7 ± 1.7
Washed roots	284 ± 274	43 ± 27
Surface sterilized roots	344 ± 262	43 ± 6
Soil B		
Rhizosphere soil	138 ± 87	0.5 ± 0.3
Surface sterilized roots	554 ± 246	88 ± 5

[a] From Döbereiner and Baldani 1979.

while the highest proportion of low-level streptomycin-resistant bacteria was found within roots.

The capability of 16 out of 24 *A. lipoferum* strains and 10 out of 21 *A. brasilense* strains to produce bacteriocins which inhibit several strains of enterobacteria, streptococci, and several unidentified Gram-negative bacteria (Oliveira and Drozdowicz 1981) may play an additional role in the competition of azospirilla within the rhizosphere microflora.

The selective enrichment of azospirilla in the rhizosphere has also been attributed to a strong chemotactic attraction of these organisms by root exudates. Azospirilla responded chemotactically to alanine, lysine, valine, aspartate, glycine, succinate, malate, maltose, and sucrose (Alvarez-Morales and Lemos-Pastrana 1980a, b; Barak *et al.* 1983; Heinrich and Hess 1985; Reinhold *et al.* 1985), even though the two disaccharides are not metabolized by this organism. In wheat roots, sucrose is cleaved by invertase to glucose and fructose which have strong chemotactic attraction for *A. lipoferum* (Hess *et al.* 1985.) There are differences between strains in terms of chemotactic response. *Azospirillum lipoferum* was attracted more strongly than *A. brasilense* by maize and sorghum exudates (Heinrich and Hess 1985) and *Azospirillum* isolates from C_4 plants showed maximal chemotactic response to malate, whereas two isolates from wheat (C_3 plant) responded preferentially to aspartate, citrate, and oxalate (Reinhold *et al.* 1985). In agreement, axenically grown C_4 plants (maize and Kallar grass) exuded predominantly malate, whereas the most abundant exudate from rice (C_3 plant) is oxalate (Kloss 1981).

Soil pH and Al^{3+} concentration also affect the occurrence of azospirilla. Earlier observations on pH requirements of *A. brasilense* (strain Sp7) (Döbereiner and Day 1976; Okon *et al.* 1977a) and on the occurrence of *Azospirillum* spp. in soils (Döbereiner *et al.* 1976) have led to the general belief that these organisms, like azotobacters, are very pH demanding. While 70% of the soils with pH values between 5.8 and 6.2 contained azospirilla, their presence was observed only in 40% of soils with pH values between 4.8 and 5.1 (Döbereiner 1978a). Roots, however, seem to provide more favorable environments for azospirilla growth even in acid soils. In Poland, azospirilla occurred in roots of various cereals grown in soils with a pH range from 5.2 to 8.2 (Kulinska 1983). Also, there was no correlation between soil pH and numbers of azospirilla in surface-sterilized roots of grasses grown in soils with a pH range of 5.8 to 7.2, while numbers in soil showed a highly significant ($r = 0.90$) correlation with soil pH in a similar range (Döbereiner *et al.* 1976). More recent observations indicate that *A. amazonense* and also most *A. lipoferum* and some *A. brasilense* strains grow as well or better at pH 6.3 (or even 5.7) than at pH 7.0. Root isolates seem especially more tolerant to lower pH values (Döbereiner 1983). Besides pH effects, Al^{3+} toxicity seems to affect the occurrence of azospirilla in soil. In three maize fields with different treatments, 36 azospirilla-positive soil samples contained 0.45 ± 0.16 mE Al^{3+}/100 g, whereas the 55 azospirilla-negative soil samples contained 1.36 ± 0.15 mE (Döbereiner 1978a).

The role of pesticides in azospirilla ecology has not been exhaustively studied. Diuron inhibited nitrogenase activity of soil inoculated with *A. brasilense* and amended with malate, while Simazine (in concentrations of up to 200 ppm) increased it (Tavares 1978). This herbicide and Atrazin were also shown to increase azotobacter counts (Bopaiah and Rai 1979). Numbers of azospirilla in soil increased also with the commercial application levels of Benomyl and 2-aminobenzimidazole (Charyulu *et al.* 1980). Bromoxynil and Afalon S increased nitrogenase activity of excised maize roots and the dry weight of maize inoculated with *A. brasilense* (Fayez *et al.* 1983). Similar observations were made with Atrazin and Alachlor, which also increased total plant N content and N percentage in maize leaves (Marriel and Cruz 1978). These promising preliminary observations, however, have not been further expanded.

The high temperature requirements of azospirilla, with optimum growth between 32 and 36°C, may explain the wide occurrence of these organisms in tropical environments. There seem, however, to be pronounced differences between strains in N_2-dependent growth at relatively low temperatures (22°C), which could be of considerable ecological importance (Neves *et al.* 1976). The existence of more cold-tolerant azospirilla was suggested by Weier *et al.* (1981) who observed active associations with *Panicum maximum* in subtropical Australia. No data are available concerning temperature optima for strains from various climatic regions, except for preliminary observations in our laboratory which indicate that some isolates from Patagonia grew better

Table 2.13 Occurrence of *Azospirillum* spp. associated with tropical forage grasses[a,b]

	Soil	Washed roots	Surface sterilized roots[c]
Total no. of samples	115	345	345
% samples containing *A. lipoferum* and/or *A. brasilense*	94	85	45
% samples containing *A. amazonense*	42	53	32
% samples containing more than 10^5 *A. brasilense* and/or *A. lipoferum*	8	34	10
% samples containing more than 10^5 *A. amazonense*	2	16	1

[a]From Magalhães *et al.* 1983.

[b]*Hyparrhenia rufa, Digitaria decumbens* cv. Transvala, *Brachiaria decumbens* cv. Pangola, *Pennisetum purpuretum* cv. Elefante, and *P. purpurem* cv. Cameron were grown in a field experiment with three N levels and with five replicates sampled monthly during 1 year. There were no apparent differences between N levels or plant genotypes.

[c]15 min. in chloramine-T (1%).

at 25°C than isolates from Brazil. Higher temperatures (above 36°C) affect growth more effectively than nitrogenase activity, which remained constant between 28 and 41°C (Day and Döbereiner 1976; Neves *et al.* 1976). At 45°C, however, nitrogenase activity was completely inactivated within 30 minutes without causing cell death (Neves *et al.* 1976). The new *Azospirillum* species, *A. halopraeferans*, isolated from Kallar grass root surfaces in Pakistan was observed to have a temperature optimum of 41°C (Reinhold *et al.* 1987).

Little is known about the ecology of individual *Azospirillum* spp. Table 2.13 shows a comparison of the occurrence of three species in soil and roots of various forage grasses and cereals in the state of Rio de Janeiro, Brazil. *Azospirillum amazonense* seems less common and also occurs in lower numbers. All three species seem to be most numerous at the root surface (washed roots). In the Amazon region, *A. amazonense* seemed to be more common in soil and roots of established pastures (33% of samples) than in nutrient depleted soils (6% of samples) (Magalhães and Döbereiner 1984). It has also been isolated from sugarcane roots in Hawaii and Thailand (J. Döbereiner, unpublished data). Also interesting is the observation that relatively high numbers of *A. amazonense* occur in roots of a palm tree (*Bractis gasipeas*), while numbers of *A. lipoferum* and *A. brasilense* were much lower here (Magalhães and Döbereiner 1984).

3

Other Root-Associated Diazotrophs

Besides azospirilla, several other diazotrophic bacteria have been found to occur in the rhizosphere and on and in roots of various plants, especially cereals. There are several facultative anaerobes (most of them *Bacillus* spp. and Enterobacteriaceae), microaerophilic organisms resembling azospirilla, and aerobes that can fix N_2 under air. In many cases these bacteria occur in much higher numbers than azospirilla and therefore may play an important role in associative N_2 fixation (Bally *et al.* 1983; Van Berkum and Bohlool 1980). In many cases, however, neither bacteria nor plant–bacteria associations have been characterized to the extent that the observations could be confirmed in other laboratories. Only well-defined groups will be described here.

3.1 Facultative anaerobic N_2-fixing bacteria

In facultative anaerobes N_2 fixation is restricted to anaerobic conditions, but growth with combined N can proceed normally in the presence of air. This is due to the lack of a protection mechanism against O_2 of the nitrogenase of facultative anaerobes. Because fermentative metabolism is much less efficient in terms of energy production than oxidative phosphorylation, the amount of N_2 fixed per gram of carbon substrate consumed is usually in the range of 1 to 6 mg, less than half of that of aerobes. The role of facultative diazotrophs in association with plants, however, has to be viewed within the microbial community where there may be associations with other diazotrophs that could use their metabolites, thus making more efficient use of available substrates. Another ecological advantage of some facultative anaerobes is the spore-forming ability of *Bacillus* that makes these organisms resistant to adverse environmental conditions. In most facultative anaerobes diazotrophy is not a criterion for the taxon, and there are strains that fix N_2 and others that do not in the same species, independently of their origin (Postgate 1982).

***Bacillus* spp.** Until recently, the only *Bacillus* species known to contain strains capable of fixing N_2 were *B. polymyxa* and *B. macerans* (Witz *et al.* 1967). *Bacillus circulans* was also suggested to be diazotrophic (Line and Loutit 1971), but this could not be confirmed by L. Seldin (personal communication). The most convincing data about plant root-associated bacilli were obtained with a "*B. polymyxa*-like" organism (Rennie and Larson 1979), which later could not be identified as such or as any other known N_2-fixing *Bacillus* in DNA homology experiments (Seldin and Dubnau 1985); its identity remains unclear. *Bacillus polymyxa* was shown to be responsible for N_2 fixation in xeric grasses in Utah (Wullstein *et al.* 1979), and *B. polymyxa* and *B. macerans* were isolated from the rhizosphere of prairie grasses in Canada (Rennie and Larson 1981).

Recently, a new, very efficient N_2-fixing *Bacillus, B. azotofixans*, was isolated from soil and grass and wheat roots in three distinct regions of Brazil and from sugarcane roots in Brazil and Hawaii (Seldin *et al.* 1984). The organism is similar to *B. polymyxa* and *B. macerans*, but it could be clearly distinguished by several physiological characteristics confirmed by DNA homology experiments (Seldin and Dubnau 1985). In contrast to those two species, it can use N_2 as the sole N source for growth, and all 16 isolates were able to fix N_2 in relatively high rates (20 to 240 nmol C_2H_4/ml/hour). Also, the efficiency of N_2 fixation compared in the same experiment was 10 times higher (4.1 ± 1 mg N/g glucose in contrast to 0.4). Most important is the capacity of *B. azotofixans* to fix N_2 in the presence of high levels of NO_3^- (50 m*M*), indicating that it does not possess any nitrate reductase activity and can therefore fix N_2 in soils fertilized with NO_3^-. Differential characteristics for the three species are presented in Table 3.1.

Enterobacteriaceae Diazotrophic strains occur naturally in several species of the genera *Enterobacter, Klebsiella, Escherichia, Erwinia*, and *Citrobacter*, but not in *Escherichia coli* and in the genera *Proteus, Serratia, Salmonella*, and *Shigella* (Postgate 1982). Species commonly found associated with the roots of plants are *Enterobacter cloaceae, E. agglomerans, Klebsiella pneumoniae, K. oxytoca, Erwinia herbicola*, and several other unidentified enterobacteria (Haahtela *et al.* 1978; Hirota *et al.* 1978; Nelson *et al.* 1976; Pedersen *et al.* 1978; Raju *et al.*

Table 3.1 Comparison of the characteristics of *Bacillus azotofixans*, *B. polymyxa*, and *B. macerans*[a]

Characteristic	*B. azotofixans*	*B. polymyxa*	*B. macerans*
% DNA homology with *B. azotofixans*	77 ± 14[b]	11	20
Mol % G + C	47.9–52.5	43.1–45.6	49–51
Nitrogen fixation	+	Variable	Variable
N_2 used as sole N source for growth	+	−	−
Spore position[c]	C–ST	C–T	ST–T
Voges–Proskauer	+	+	−
Acid and gas from xylose, arabinose, lactose, ribose, glycerol	−	+	+
Growth at 45°C	−	−	+
Nitrate reduction	−	+	+
Nitrogenase activity with 50 mM NO_3^-	+	−	−
Starch hydrolysis	−	+	+
Decomposition of casein	−	+	−
Liquefaction of gelatin	−	+	
Decomposition of pectin	−	+	+
Sensitivity to phages Epy-1, Spy-1, Spy-2	−	+	−

[a]From Seldin *et al.* 1984, and Seldin and Dubnau 1985. Symbols: +, positive; −, negative.

[b]Mean with standard deviation of 16 strains.

[c]Spore position: C–ST, central to subterminal; C–T, central to terminal; ST–T, subterminal to terminal.

1972; Rennie 1980a; Singh *et al.* 1983). Most of these bacteria occur in large numbers on and in roots and usually outnumber azospirilla (Bally *et al.* 1983; Haahtela *et al.* 1981; Klucas *et al.* 1981; Kulasooriya *et al.* 1981; Lindberg and Granhall 1984; Rennie *et al.* 1982; Wright and Weaver 1981). The high instability of diazotrophy, observed during the purification of putative diazotrophs from mixed populations, suggests that syntrophic interactions among the various bacterial species were needed for expression of nitrogen fixation or that unstable *nif* plasmids were lost under nonselective conditions (Lindberg and Granhall 1984). The latter applies to all diazotrophs whose *nif* genes are plasmid borne, and it is possible that a plant factor, initially present in isolates but lost in culture, is required for *nif* expression. A situation possibly analogous to that observed with fast-growing *Rhizobium* is their failure to express *nif ex-planta*. Plasmid-borne *nif* genes have up to now been found only in *Enterobacter agglomerans* (Singh *et al.* 1983). In *K. pneumoniae* strain M5a1 and *K. oxytoca* strain NG13, *nif* genes are chromosomal (see Postgate 1982; Wang *et al.* 1985). The ability of P^- and F^- incompatibility group plasmids in mobilizing and transferring the entire cluster of *nif* genes of *K. pneumoniae* M5a1 to several nondiazotrophic enterobacteria (Dixon and Postgate 1972; Kleeberger and Klingmüller 1980; Krishnapillai and Postgate 1980; Postgate and Krishnapillai 1977; Postgate and Kent 1985) has been demonstrated. This fact and the presence of naturally occurring *nif* plasmids, their transmissibility, and the demonstration of suitable genetic background for their expression in most genera of Enterobacteriaceae (Postgate 1982; Postgate and Kent 1985; Singh *et al.* 1983) could explain the observed scattering of diazotrophy in the various species of this group of bacteria.

According to Postgate (1982), *Klebsiella pneumoniae* strain M5a1, being the best-studied of the free-living diazotrophs, is actually a strain of *K. oxytoca*. Diazotrophic strains of *K. oxytoca* have been found in the rhizosphere of rice (Thomas-Bauzon *et al.* 1982; Wang *et al.* 1985). Cloning and restriction enzyme analysis of *nif* genes from *K. oxytoca* NG13, a strain reputed as an efficient N_2 fixer associated with rice (Hirota *et al.* 1978), showed that it had a high degree of base sequence similarity with that of analogous genes of *K. pneumoniae* strain M5a1 (Wang *et al.* 1985). These observations suggest that *K. pneumoniae* M5a1, a free-living diazotroph of no known agronomic importance, may turn out to be an associative diazotroph of yet unknown potential. Its close association with plant roots, however, remains to be demonstrated. Few data are available on plant interactions with enterobacteria. Haahtela and Korhonen (1985) observed *in vivo* adhesion of *Klebsiella* to plant roots. A positive correlation between numbers of *E. cloaceae* and *K. pneumoniae* with nitrogenase activity of intact root soil cores of forage grasses examined in Texas has been reported (Wright and Weaver 1981).

Enterobacteriaceae comprise anaerobes and facultative anaerobes. Facultative anaerobes, such as *K. pneumoniae* strain M5a1, fix nitrogen anaerobically, but also fix nitrogen under conditions of oxygen limitation. Oxygen inactivates preformed nitrogenase and represses its synthesis (Eady *et al.* 1978), but the introduction of low O_2 concentrations to previously anaerobic chemostat cultures increases the molar growth yield and the efficiency of nitrogen fixation (Hill 1976). Stimulation of nitrogen fixation by oxygen is believed to be associated with a general upgrading of the energy status of the cell, supported by oxidative phosphorylation (Hill 1985). Oxygen affects nitrogenase activity and synthesis in different ways. Inactivation of nitrogenase and repression of transcription of all *nif* genes, except *nif*LA, occurs at about 6 μM O_2. Higher levels of O_2 (60 μM) are needed to decrease expression of the *nif*LA operon (Hill 1985). Although associated with the anaerobic metabolism of facultative anaerobes, expression of diazotrophy in these organisms does not require the participation of the *fnr* gene product, a product whose activity is needed for expression of anaerobic catabolism in Enterobacteriaceae (Hill 1985).

A large number of mono- and disaccharides and polyols support growth and nitrogen fixation by *Klebsiella* and *Enterobacter*. Sucrose was found to be the best carbon source followed (in order) by D-fructose, D-glucose, and D-mannitol. Among the organic acids tested (fumarate, malate, lactate, α-ketoglutarate, succinate, and oxaloacetate), only the first two were used (Haahtela *et al.* 1983b).

The effect of phosphate concentration was studied on diazotrophically and nondiazotrophically grown *K. pneumoniae* strain M5a1 (Bergersen 1974). The results did not support the previous belief that N_2-fixing cells had a higher requirement for phosphate and indicated that the primary effect of phosphate limitation was to restrict cellular biosynthetic processes (Bergersen 1974).

Amino acids and NH_4^+ under certain conditions repress nitrogenase synthesis in *K. pneumoniae* and other Enterobacteriaceae, as in other diazotrophs, with the exception of *Rhizobium* (see Dixon 1984; Eady, 1981; Eady *et al.* 1978).

3.2 Microaerobic N_2-fixing bacteria

This rapidly growing new group of diazotrophs resembles azospirilla especially with respect to the requirement of low O_2 concentrations for nitrogenase activity. These organisms possess a respiratory metabolism and grow rapidly under air with combined N but not with N_2. The use of semisolid media introduced for the isolation of azospirilla seemed to be the key for the discovery of all of these new diazotrophs that occur preferentially in the plant root environment.

***Herbaspirillum* spp.** During azospirilla counts carried out on cereal roots in the State of Rio de Janeiro, Brazil (Baldani 1984), the frequent occurrence of a smaller vibrioid organism was observed which in N-free semisolid media formed pellicles similar to those of azospirilla. A large number of isolates (120) with very homogeneous characteristics were recovered from rhizosphere and root samples of maize, sorghum, and rice in the State of Rio de Janeiro and from sugarcane in Hawaii. According to *Bergey's Manual of Systematic Bacteriology* (Krieg and

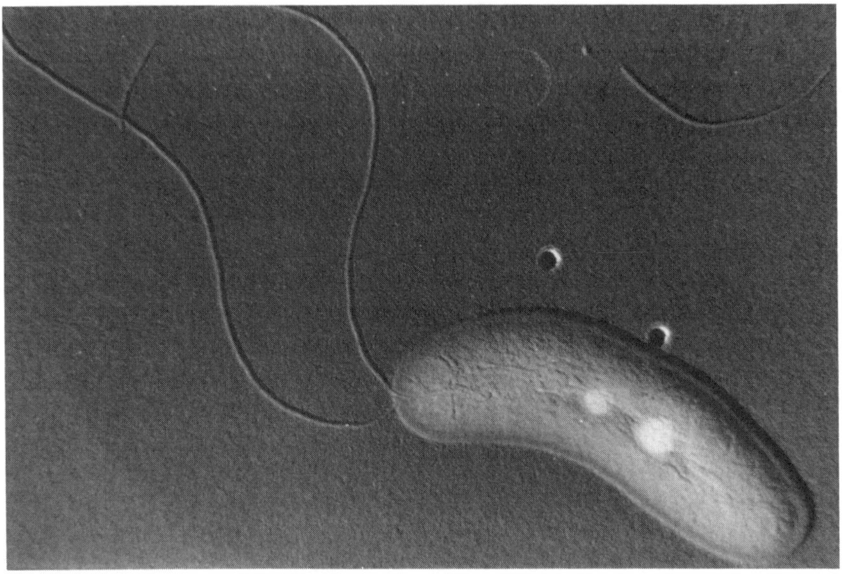

Figure 3.1 *Herbaspirillum seropedicae* showing the usual double, polar flagella. Single, polar or bipolar flagella are also occasionally observed (14,000×). (Courtesy N. R. Krieg.)

Holt 1984), this organism was classified within Section 2, where due to the G + C content of DNA, it was provisionally placed into the genus *Azospirillum* (Baldani *et al.* 1984). However, based on more detailed studies with emphasis on recent RNA/RNA homology studies (Falk *et al.* 1986), a new genus with one species was proposed (Baldani *et al.* 1986c).

The description of the species according to these authors is as follows. *Herbaspirillum seropedicae* gen. nov. sp. nov. (Baldani, Baldani, Seldin, and Döbereiner 1986c) (grass root-associated bacteria first isolated in Seropédica, State of Rio de Janeiro, Brazil). Cells are Gram-negative, generally vibrioid, sometimes helical, and very motile when exposed to O_2. Two polar flagella, occasionally with one to three flagella on both poles (Figure 3.1). Cell diameter is 0.6 to 0.7 μm, and cell length varies with the medium from 1.5 to 5 μm. On soft nutrient agar at 35°C there is pronounced swarming, but lateral flagella have not been observed. *Herbaspirillum seropedicae* has a typical respiratory metabolism and sugars are not fermented. It fixes atmospheric N_2 under microaerobic conditions and grows well with N_2 as the sole nitrogen source. Efficiency is 12 to 15 mg N_2 fixed/g DL-malate. Growth on N_2 is somewhat slower than that of azospirilla even though growth on mineral nitrogen or glutamate is much faster. It is oxidase, catalase, and urease positive and does not grow in 2% NaCl. Organic acids such as malate, fumarate, succinate, pyruvate, citrate, and *trans*-aconitate are the favored carbon substrates both for NH_4^+- and N_2-dependent growth. Mannitol, sorbitol, and glycerol and also several sugars (glucose, galactose, and L-arabinose) are oxidized. Starch and gelatin are not hydrolyzed. Vitamins or growth substances are not required. It is sensitive to chloramphenicol, tetracyclin, gentamycin, kanamycin, erythromycin, and streptomycin. Optimum growth temperature is 34°C with no growth beyond the limits of 22 and 38°C. The optimum pH for N_2-dependent growth seems broader than for azospirilla, and equally good growth is observed between pH 5.3 and 8.0. Nitrate is assimilated and/or dissimilated to NO_2^- under O_2 limitation. No NO_3^--dependent anaerobic growth and no visible gas from NO_3^- are observed in solid or semisolid media. Small amounts of N_2O are formed under C_2H_2. The mole percent G + C content of the DNA is 67 ± 0.5. Its habitat is soil and Gramineae roots. Strain Z67 has been designated as the type strain (ATCC 35892) and strains Z78 (ATCC 35893) and Z152 (ATCC 35894) represent additional isolates.

A comparison of *Herbaspirillum* with *Azospirillum* and other diazotrophs is presented in Table 2.3. In comparison with *Azospirillum* spp., the most prominent differences of the new genus are the smaller cell size, the presence of more than one flagellum, the distinct colony type, and the slightly lower mole percent G + C of its DNA. Other differences are the broader pH tolerance and the higher tolerance of nitrogenase activity to O_2. The quantification of DNA homology, shown in Table 2.2, confirms the difference between *H. seropedicae* and the three *Azospirillum* species.

Recent results of RNA/RNA hybridization experiments comparing *H. seropedicae* with the three *Azospirillum* spp. and several other possibly related N_2-fixing bacteria (Table 2.2) confirmed very low RNA homology (less than 20%) (Falk *et al.* 1986). *Herbaspirillum seropedicae* therefore can not be included in the genus *Azospirillum*. The closest related organisms seem to be some *Aquaspirillum* spp., especially *A. itersonii* and *A. peregrinum*, and *Rhodospirillum* spp. (Table 2.2). All three contain strains that can fix N_2 under microaerobic conditions. *Herbaspirillum seropedicae*, however, can be distinguished by the vibrioid cell shape and its double flagella, usually on one pole, and by its inability to grow photoautotrophically, its capacity to use several sugars, and its occurrence in soil and roots in contrast to *Aquaspirillum* and *Rhodospirillum* which are typical freshwater organisms. Furthermore, *A. itersonii* does not contain urease. Coccoid bodies, which are formed in old cultures of *Aquaspirillum* species and water-soluble pigments that diffuse into the medium, have not been observed in *Herbaspirillum*. RNA/RNA hybridization experiments confirmed the very remote relationship of *Herbaspirillum* with *Rhodospirillum* and *Aquaspirillum* (Falk *et al.* 1986).

***Pseudomonas* spp.** Pseudomonads are very common in the rhizosphere (Vancura 1980), and inoculated strains appeared to establish themselves well, representing 25% of all pseudomonads in the rhizosphere of wheat (Weller 1984). This competitive ability has been attributed to the production of metabolites that are inhibitory to microbial growth (Lesinger and Margroff 1979). Some *Pseudomonas* strains have been found to be able to fix N_2 (Bally *et al.* 1983; Barraquio *et al.* 1983; Haahtela *et al.* 1983a) and have been suggested to be important in association with cereals. Due to the imprecise description of the genus (Holt and Krieg 1984), identification of species is unclear, and therefore the confirmation of results from various laboratories of *Pseudomonas*-cereal associations is difficult.

Bally *et al.* (1983) identified one group of their isolates of *P. paucimobilis* as the most common among three *Pseudomonas* spp. in the rice rhizosphere. Barraquio *et al.* (1983) isolated many strains of a N_2-fixing *Pseudomonas* sp. which they considered to be a new species, but no name was assigned to it. These organisms were obtained from the root surface, from within roots (histosphere) of rice and wild rice (*Oryza punctata* and *O. australiensis*), and from an aquatic weed (*Monochoria vaginalis*) growing in rice fields (Barraquio and Watanabe 1981). They were isolated from soil or root dilutions (10^{-5} to 10^{-6}) enriched in a semisolid glucose medium containing yeast extract. These were then streaked out on tryptic soy agar where they can be easily recognized as small, white, glistening colonies that later turn grayish. A diffusible brown pigment is formed around the colonies in yeast mannitol agar. When transferred into a semisolid glucose medium containing yeast extract, a white pellicle is formed 1 to 3 mm below the surface and which reduces C_2H_2 at relatively low rates (about 10 μmol/tube/hour).

The organism is unable to grow on N_2 as the sole N source, and the capacity to fix N_2 is frequently lost in stock cultures. Dinitrogen fixation has been confirmed with the incorporation of $^{15}N_2$ (0.25 atom percent excess in comparison with 1.16 in *A. lipoferum* tested in the same experiment) (Watanabe and Barraquio 1979). This *Pseudomonas* sp., according to Barraquio *et al.* (1983), is a small (0.7×2.3 μm) Gram-negative rod occurring singly or in pairs with one polar flagellum, but motility is seldom observed. Poly-β-hydroxybutyrate is formed as carbon reserve. It uses most sugars and also organic acids, with acid formation from glucose. NO_3^- is reduced to NO_2^- but is not further denitrified. The bacteria are facultatively chemolithotrophic in an atmosphere containing H_2, O_2, and CO_2. In fact, 95% of all H_2-oxidizing bacteria that predominate in paddy rice roots were N_2-fixing species (Watanabe *et al.* 1982a). The mole percent G + C of the DNA of this *Pseudomonas* sp. is 63.8.

The *Pseudomonas* strain isolated from grass roots by Haahtela *et al.* (1983a) seems to be able to grow on N_2 as the sole N source when incubated in semisolid media under reduced O_2 tensions. The cells are short rods (0.8 × 1.5 μm) with four to ten (usually six) polar flagella. Yellow-green fluorescent pigmentation is observed in King B agar (King *et al.* 1954). Low rates of nitrogenase activity and the frequent loss of the capacity of N_2 fixation are common as reported for the *Pseudomonas* sp. described by Barraquio *et al.* (1983). Such characteristics, including the inability to grow with N_2 as the sole N source, do not disqualify these organisms as potential root associates if it is taken into account that rhizobia have similar characteristics in culture.

Other microaerobic diazotrophs Several additional diazotrophs which resemble azospirilla were isolated in N-free semisolid malate media. Only the three best characterized ones are summarized here.

Isolates obtained from surface sterilized roots of various grasses in Israel and Turkey (Nur *et al.* 1980b) consist of curved rods with one polar flagellum that are smaller than azospirilla (0.5 to 0.6 × 3 to 4 μm) and that form yellow colonies on potato agar. Growth in semisolid malate medium is similar to that of azospirilla but requires yeast extract. Nitrate is reduced to NO_2^- and further to N_2, but anaerobic NO_2^--dependent growth has not been observed.

Another microaerobic diazotroph requiring yeast extract was isolated from washed roots of *Digitaria sanguinalis* (crab grass) in Oregon (Barber and Evans 1976). The organism is still smaller (0.5 × 1 μm) and is a Gram-negative rod with a single polar flagellum. Nitrogenase activity in liquid medium supplemented with yeast extract incubated under low p_{O_2} reached 6 nmol C_2H_4/ml/hour. The mole percent G + C of its DNA is close to that of azospirilla (68%).

A small obligate microaerobic diazotroph has been found in association with cord grass roots (*Spartina alterniflora*, a seaweed). The organism resembles azospirilla with slender curved rods motile by a single polar flagellum. But because of its small size (0.3 to 0.6 × 1 to 3 μm), the low G + C mole percent of its DNA (32%), and several other characteristics, it was identified as a *Campylobacter* sp. (McClung and Patriquin 1980). This organism could be isolated from extracts from surface-sterilized roots incubated in semisolid malate medium containing yeast extract, plated on melted soft agar (7 g/liter) and incubated under low p_{O2}. Small, white, irregular colonies were transferred again into semisolid media and checked for C_2H_2 reduction.The organism uses organic acids but not sugars and requires high concentrations of NaCl (1 to 4%).

3.3 Aerobic N_2-fixing bacteria

Although most observations on root-associated bacteria deal with microaerobic organisms, which like rhizobia can fix N_2 only under low dissolved O_2 concentrations, there are some earlier reports on selective multiplication and establishment of true aerobic diazotrophs in the rhizosphere soil of Gramineae and other plants, but not in roots. These bacteria, in contrast to the former group, can fix N_2 and can grow under air since they possess respiratory protection mechanisms for their nitrogenases (see Section 2.3 on nitrogen metabolism). A comparison of these organisms is presented in Table 3.2.

***Azotobacter* spp.** The classical aerobic "free-living" N_2-fixing bacteria are *Azotobacter* spp. These organisms can be readily isolated by plating soil dilutions on N-free agar plates where they form milky white, fast-growing, smooth colonies; other bacteria cannot grow except for a few nitrogen-scavenging organisms that form translucent or water-drop-like colonies. Growth in semisolid media is on the surface where thick pellicles are formed. Further characteristics of the three *Azotobacter* spp. that occur in the rhizosphere are summarized in Table 3.1. *Azotobacter* cells are typically large plump rods often occurring in pairs. Cysts are not resistant to high temperatures, but they are more tolerant than vegetative cells to UV and γ radiation, sonic treatments, and especially to desiccation (Socolofsky and Wyss 1962). For more detailed information on the physiology and genetics of *Azotobacter* spp., see Yates and Eady (1979).

The occurrence of *A. chroococcum* and *A. vinelandii* is limited to soils with almost neutral pH, and thus they occur only sporadically in humid tropical soils (Table 3.3). In cultivated soils of temperate regions azotobacters usually are found in low numbers (Vancura *et al.* 1959) and frequently are suppressed in the rhizosphere and especially on the root surface (Brown *et al.* 1962). Multiplication of *A. chrooccoccum* in the outer rhizosphere has been observed only in the seedling stage (Jackson and Brown 1966; Rovira 1963). Vancura *et al.* (1959) could increase *Azotobacter* numbers on roots of barley, but not in rhizosphere soil, by inoculation with certain "adapted" strains but even so numbers were lower on roots than in rhizosphere soil. The poor performance of azotobacters in the rhizosphere was attributed to accumulation of bacteriostatic substrates (Barea *et al.* 1978; Strzelczykova and Strzelczyk 1961) and to the lack of usable substrates (Rovira 1965).

Table 3.2 Comparison of aerobic N_2-fixing bacteria that associate with plant roots[a]

Characteristic	*Azotobacter chroococcum, A. vinelandii*[b]	*Azotobacter paspali*	*Beijerinckia indica*[b]	*Beijerinckia fluminensis*	*Derxia gummosa*[b]
Cell size (in μm)	2–3 × 3–6	1.2 × 4–10	1–2 × 1–4	1–2 × 3–4	1 × 3–6
Pleomorphism	−	Long filaments	−	−	+
Cysts	+	+	−	Zoogloeia	−
Peritrichous flagella	+	+	+	+	−
Polar flagella	−	−	−	−	+
Aerobic growth on N-free media	+	+	+	+	±[c]
N_2 fixation in the presence of NO_3^-	−	+	±	±	−
H_2O-soluble pigments	+	+	−	−	−
Growth in N-free liquid medium	Smooth white or yellow	Flocculated yellow or purple[d]	Gummy white	Snow white	Gummy beige
Acid production	−	+	+	+	+
Use of sucrose	+	+	+	+	+
Autotrophic growth with H_2 or CH_4	−	±	±	±	+
Optimal growth (°C)	20–32	35–37	25–32	26–33	25–37
pH range for growth	5.5–8.0	5.5–8.5	3.0–10	3.5–9.2	5.5–9.0
Mol % G + C of DNA	63–67	67	55–59	56	69–73

[a]From Holt and Krieg 1984.

[b]Other species of these genera are known but have not been observed in association with plants.

[c]Only if cell clumps are in the inoculum which permit start of N_2-dependent growth.

[d]Yellow pigment and flocculated growth in stagnant broth and purple pigment with smooth growth in shaken culture with Ca citrate.

The only reports of abundant occurrence of these two species are from warm arid regions (Abd-el-Malek 1971) where *Azotobacter* enrichment in the outer rhizosphere of wheat, cotton, and legumes was observed (Vancura *et al.* 1965) and where numbers as high as 10^8 were reached (Abd-el-Malek 1971). Xerophytic plants could also promote an increase in *Azotobacter* numbers from about 10 in desert soils to 10^6 in the rhizosphere (Wullstein *et al.* 1979). An aerobic diazotroph tentatively identified as *A. chroococcum* has been isolated from root pieces of *Eragrotis ferruginea* in Japan (Nioh 1979) and from maize (Martinez-Toledo *et al.* 1985).

Azotobacter paspali was added to the genus in 1966 (Döbereiner 1966), being distinguished from the classic species by its slower growth (colonies on N-free media appear after 48 hours in contrast to 24 hours), its long filaments (up to 60 μm) after 48 hours growth in N-free liquid media, acid production with sugars, and several other characteristics (Table 3.2). Interesting is its ability to fix N_2 at high NO_3^- levels (Peña and Döbereiner 1974). Its most remarkable characteristic is the almost absolute host–plant specificity to one ecotype (Batatais) of *Paspalum notatum*, a subtropical invasion grass. Among 252 root surface samples from this ecotype collected throughout Brazil and in the United States, Argentina, Paraguay, and South Africa, 98% contained this species while only a few samples of other *P. notatum* ecotypes or other *Paspalum* spp. contained the organism. No other plant contained the bacteria (Table 3.3).

Table 3.3 Occurrence of aerobic N_2-fixing bacteria on the root surface of forage grasses in Brazil[a]

	No. of samples	% positive: *A. chroococcum*, *A. vinelandii*	*A. paspali*	*Derxia gummosa*
Paspalum notatum cv. Batatais[b]	76	17	98	15
P. plicatum	3	33	66	0
Paspalum (other species)	27	7	0	0
Other Gramineae	63	26	0	0
Leguminosae	12	16	0	8
Other families	10	4	0	20

[a]From Döbereiner 1966.

[b]Additional surveys of the occurrence of *A. paspali* (Döbereiner 1970) confirmed its presence in all but 4 of 252 root surface soil samples and its absence in 200 samples from other plants from Brazil, the United States, Paraguay, Argentina, and South Africa.

The high pH requirements of the organism seem to be satisfied in the rhizosphere and especially on the root surface, where the pH of *P. notatum* Batatais is higher than that of the soil and where abundant *A. paspali* occurrence was reported in soils with a pH of about 5.0 (Döbereiner 1966).

***Beijerinckia* spp.** In contrast to *Azotobacter, Beijerinckia* spp. are known as tropical representatives of aerobic diazotrophs. *Beijerinckia indica* was first isolated from an acid Indian soil cultivated with rice and named *Azotobacter indicum* (Starkey and De 1939), but was later renamed on the basis of a larger number of isolates that contained two species, namely, *Beijerinckia indica* and *B. mobile* (Derx 1950). Two other species were added later and named *B. derxii* (Tchan 1957) and *B. fluminensis* (Döbereiner and Ruschel 1958).

Beijerinckia spp. form characteristic slow-growing colonies on N-free silica gel or agar media and appear only after 8 to 10 days. *Beijerinckia indica* colonies are very characteristic; they are tough and sticky or even elastic and dense, white at the beginning and later buff-colored with a characteristic smell. Liquid media become viscous throughout. Colonies of *B. fluminensis* grow even slower and are dry and look like sand humps. Colonies of the other species are less characteristic, and perhaps for this reason their ecological distribution has not been further studied. Physiological characteristics of *Beijerinckia* spp. are summarized in Table 3.2. Maximal efficiency in N_2 fixation was reported as 31 and 13 mg N/g sucrose for *B. indica* and *B. fluminensis*, respectively (Döbereiner 1968).

The most characteristic feature of all *Beijerinckia* spp. are the two distinct polar poly-β-hydroxybutyrate globules that are readily seen at both extremities of the cells and can become so large that they fill almost the whole cell, especially in older cultures. *Beijerinckia* spp. are very acid tolerant, growing in media with pH values of 3 to 9; they have been reported to occur in high numbers in many tropical soils. In a survey of 392 soil samples collected worldwide, *Beijerinckia* occurred only rarely in temperate or subtropical regions but was found abundantly in tropical areas of Africa, Southeast Asia, and South America (Becking 1959; Döbereiner and Castro 1955). The preferential occurrence of the organism in tropical soils was originally attributed to its acid tolerance (Becking 1961a), but was later suggested to be due to a preference for lateritic soils (Becking 1961b).

Both species have been found to be selectively enriched in the rhizosphere and especially on the root surface of sugarcane where other soil organisms are less common (Döbereiner 1961). Among 131 soil samples from sugarcane fields in seven different states in Brazil, 95% of them contained *Beijerinckia*, while only 60% of those bearing other

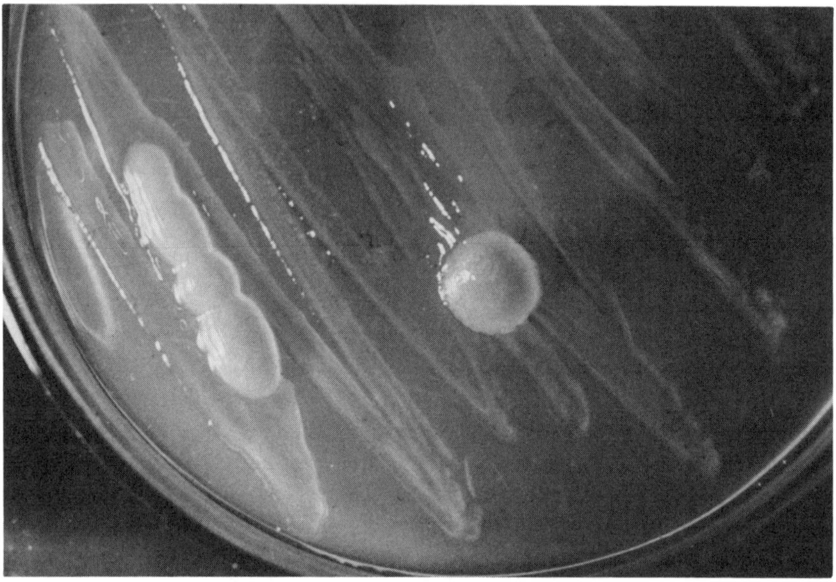

Figure 3.2 Characteristic N-free starch agar plate streaked with *Derxia gummosa*, showing transparent, thin growth in most areas and only a few N_2-fixing colonies.

crops contained the organism. *Azotobacter* was found in only 9 and 6% of these soil samples, respectively (Döbereiner 1959). There was also an increase in the number of *Beijerinckia* in the rhizosphere of various forage grasses (Ruschel and Döbereiner 1965; Souto and Döbereiner 1971) and on cotton leaves (Murty 1984).

***Derxia* spp.** Although *Derxia* is considered an obligate aerobe and can fix N_2 under air (Table 3.2), it is unable to do so if only a few cells are inoculated (Hill 1971). Growth on N-free agar plates only starts where clumps of organisms are present, which protect themselves against O_2 within their very tenacious slime. Once colonies start they grow to a very large size (20 mm in diameter and 10 mm in height) and become characteristically brown, wrinkled, and gelatinous, thus permitting their ready identification. This leads to a typical pattern, seen in Figure 3.2 of a few large brown colonies among many small translucent colonies that do not fix N_2. On media with combined N, *Derxia* grows uniformly in massive brown colonies. Several strains were found to grow autotrophically using H_2 as their sole energy source (Pedrosa *et al.* 1980). A comparison of this organism with other aerobic diazotrophs is shown in Table 3.2. Efficiency of N_2 fixation reached 25 mg N/g glucose (Jensen *et al.* 1960).

Derxia occurs mainly in tropical soils. It was originally isolated from a soil in West Bengal (Jensen *et al.* 1960) and is widely distributed in Brazil (Campêlo and Döbereiner 1970), Indonesia, China, and South Africa (Becking 1959). On a selective N-free mineral starch medium containing bicarbonate it was more often isolated from grass roots than from soil (Campêlo and Döbereiner 1970; Döbereiner 1968), but no further information on associations with plants is available.

4

Association with Plants

The isolation of a certain kind of diazotroph from the rhizosphere or even from plant roots gives little information about the nature of the diazotroph's relationship with this plant. Some specific kind of plant–bacteria interactions must exist in order to show a mutual relationship that can be called association. If there are no visible morphological structures such as legume nodules, the identification of the mechanism involved in the establishment of such associations is difficult, especially if one considers the large variability of these apparently primitive systems. Associations of azospirilla or other diazotrophic bacteria with cereals and forage grasses of this kind have been described by many authors (e.g., Boddy and Döbereiner 1984; Dart 1986; Döbereiner and De-Polli 1980; Hubbell and Gaskins 1984; Okon 1982; Van Berkum and Bohlool 1980), but the exact nature of these associations and the contribution of each partner (bacteria and plant) have so far not been defined. Even so, considerable progress has been made in recent years with the utilization of antibiotic-resistant *Azospirillum* strains that can be established within roots under field conditions, recovered, and identified (Baldani *et al.* 1986a; Marocco *et al.* 1983), and chances of faster progress in the near future are good.

4.1 Rhizosphere colonization

Colonization of plant root surfaces by rhizobacteria must occur before subsequent effects of the bacteria on plant growth can take place (Douglas *et al.* 1985). A strategy to identify inoculant bacteria for colonization or for biological control must select not only for bacteria that are capable of efficient N_2 fixation or of antagonism against plant pathogens, but also for the ability to effectively colonize roots. The screening of various soil and rhizosphere bacteria for the ability to colonize plant roots involves their introduction onto seeds (or soils) and the sampling of the root system for bacterial presence and number at intervals during the whole plant cycle.

The definition of the term rhizosphere varies widely, but in most cases the soil environment that is under the immediate influence of the root system is considered to be the rhizosphere. Root effects are primarily due to the delivery of carbon substrates for microbial growth, and therefore a gradient effect should be expected that diminishes with the distance from the root. Also pH, CO_2, and O_2 variations can be expected to occur under the influence of the root system (Marschner 1985). The region within root tissues has been called the histosphere (Watanabe *et al.* 1979). The soil layer directly adjacent to the roots is usually called the rhizoplane. Patriquin and Döbereiner (1978) proposed the terms ecto- and endorhizosphere for root surface and histosphere and several other terms are in use.

The rhizosphere effect has been known for a long time and has been considered responsible for the selective increases of microorganisms (Macura 1966; Rovira 1965, 1969) promoted by the enrichment of the soil with specific carbon substrates made available by the roots. Enrichment of diazotrophs in relation to the total microflora in the rhizosphere, as compared to nonrhizosphere soil, are not so generalized. The ratio of the number of diazotrophs in rhizosphere soil to those in nonrhizosphere soil (R/S) calculated by Balandreau *et al.* (1978) from various sources is usually in the range from 1 to 20 and is seldom more than 100. These ratios are similar to those obtained with total bacterial numbers and do not indicate selective enrichment of diazotrophs. Among the few data showing changes of the microbial equilibrium in favor of diazotrophs are those of *Beijerinckia* in the sugarcane rhizosphere (Table 4.1) and the data of Patriquin and McClung (1978) which showed R/S values of 7100 and 380 for facultative anaerobes and microaerobes, respectively, as compared to R/S values of 2.6 for total bacteria in *Spartina alterniflora*.

Selective enrichment for diazotrophs in the rhizosphere or rhizoplane might be attributable to (a) their competitive advantages in a C-rich, N-poor environment, (b) the provision by the plant of essential vitamins such as biotin (Martin and Glatzle 1982), (c) the chemotactic attraction by root exudates (Balandreau 1986), or (d) the aerotactic attraction to reduced p_{O2} in the root region (Okon *et al.* 1980).

Expression of the number of diazotrophs as a percentage of the total number of bacteria, determined either by testing isolates obtained in-

Table 4.1 Selective enrichment of *Beijerinckia* in the rhizosphere of sugarcane in the field[a]

	R/S[b]	RP/S[c]
Beijerinckia spp.	22.3	55.8
Total bacteria	0.4	0.2
Actinomycetes	1.0	0.1
Fungi	0.7	0.7

[a]From Döbereiner 1961.
[b]Ratio of number in rhizosphere soil to number in soil between rows.
[c]Ratio of number in rhizosplane soil to number in soil between rows.

itially on a nutrient agar medium for acetylene-reduction activity (ARA), or by comparison of most probable numbers of diazotrophs with separate counts of heterotrophic bacteria, has been used to illustrate the preferential enrichment of microaerophilic diazotrophs on the root surface and/or inside roots of several dry land cereals (maize, sorghum, wheat, and rice) (Baldani 1984), and even in water plants such as rice (Barraquio *et al.* 1982; Watanabe and Barraquio 1979) and *Spartina alterniflora* (Boyle and Patriquin 1980; Patriquin and McClung 1978). Watanabe and Barraquio (1979) reported that 2.4, 37.5, 76, and 81% of the bacteria isolated from rhizosphere soil, stems, rhizoplane, and histosphere, respectively, of rice exhibited acetylene-reduction activity (ARA) in semisolid media; most of these microaerophils were identified as *Pseudomonas* spp. (Table 4.2) (Barraquio *et al.* 1983). Similarly, 0.2 ± 0.15, 7 ± 5, and 85 ± 20% of the total bacterial number in soil, rhizosphere, and endorhizosphere of Kallar grass were diazotrophs (Reinhold *et al.* 1985b). Azospirilla predominated on the root surface, while unidentified motile straight rods predominated within roots. Lower percentages (1 to 10%) were observed for microaerophiles in *Spartina alterniflora* (Boyle and Patriquin 1980; Patriquin and McClung 1978). The number of azospirilla recovered from inoculated maize growing in soil was equivalent to 1% or less of the total number counted separately (O'Hara *et al.* 1981; Okon *et al.* 1977).

Except for systems using solution or semisolid plant cultures, colonization of roots by bacteria tends to be very patchy (Rovira 1965). Acetylene-reduction activity of individual root pieces from *Digitaria* (Döbereiner and Day 1976) and *Spartina* (Patriquin and McClung 1978) was found to be highly variable, and a much greater fraction of the population of microaerobic diazotrophs was found in active root segments than in the roots as a whole (Döbereiner and Day 1976).

Preferential enrichment of diazotrophs on and in the root in comparison with rhizosphere soil may reflect a continuation of gradients in

Table 4.2 Frequency of N_2-fixing microaerobic bacteria in rice and other plants[a,b]

Plant source	All N_2-fixing microaerobes[c]		N_2-fixing *Pseudomonas* spp.[d]	
	Soil	Roots[c]	Soil	Roots
Paddy rice (3 cultivars, 2 soils)	40 ± 17	56 ± 29	7 ± 10	44 ± 27
Wild rice	9	75	3	40
Aquatic weed	—	50	—	45
Dry rice (3 cultivars)	—	26 ± 5	—	0
Maize (2 cultivars)	—	2	—	0

[a]From Barraquio and Watanabe 1981.

[b]Expressed as % of total microaerobic heterotrophs.

[c]Soil or washed root homogenate dilutions were spread on tryptic soy agar and 100 colonies replicated into semisolid glucose medium with 0.01% yeast extract for C_2H_2 reduction tests.

[d]Recognized by colony type.

the factors suggested above to account for preferential enrichment in rhizosphere soil and/or involvement of "attachment factors" or specialized ability to penetrate the roots.

The studies of Shimschick and Hebert (1979) and Umali-Garcia *et al.* (1980) suggested that differences in binding of bacteria to roots could contribute to preferential enrichment of certain organisms on the rhizoplane. Shimschick and Hebert (1979) reported that rhizobia exhibit polar attachment to roots of wheat and rice. Attachment tends to occur in a monolayer, the numbers bound are time dependent and proportional to the concentration of bacteria in the medium, and there is a slow time-dependent dissociation. Umali-Garcia *et al.* (1980) compared numbers of adsorbed cells per 200-μm lengths of root hair on pearl millet incubated with standardized concentrations of various bacteria. The numbers were highest for three strains of *Azospirillum brasilense* (24 to 32); low for *Azotobacter vinelandii, Klebsiella pneumoniae*, and *Escherichia coli* (less than 0.7), and intermediate for *Pseudomonas fluorescens* and *Rhizobium trifolii* (1 to 14), suggesting some selective advantage for association with *Azospirillum brasilense. In vitro* adhesion of N_2-fixing *Klebsiella* spp. to cereal roots was observed by Haahtela and Korhonen (1985), and fimbriae (filamentous proteinous appendages on bacterial cell surfaces) seem to be involved in the attachment of N_2-fixing Enterobacteriaceae to grass and cereal root hairs (Korhonen *et al.* 1986).

There appear to be fundamental differences between diazotrophs in the processes involved in attachment. Shimschick and Hebert (1979) observed that the number of *Azotobacter* binding to cereal roots was proportional to the concentration in the medium, but unlike attachment of *Rhizobium*, binding occurred in both monolayers and multilayers and was not inhibited by streptomycin. Binding of rhizobia, in contrast, was inhibited by streptomycin, heating, or exposure to $HgCl_2$, but not by the heating of roots, suggesting involvement of a thermolabile protein of bacterial origin in the attachment process. Like rhizobia, azospirilla exhibit polar single-cell attachment to root hairs, epidermal cells, and mucigel, but also attach in multicellular clumps. Bacteria not removed by washing are generally those exhibiting polar attachment (Umali-Garcia *et al.* 1978). *Azospirillum* attaching in polar fashion to roots exhibit vigorous corkscrew motion. After several hours, fibrillar structures appear to anchor the organisms to the root (Patriquin 1981).

In their study of colonization of roots of pearl millet and guinea grass by *Azospirillum brasilense* strain Sp7, Umali-Garcia *et al.* (1980) noted a number of characteristics that, in addition to polar attachment, are believed to be involved in the *Rhizobium* attachment recognition process (Dazzo 1980). These were: (1) attachment was influenced by the age of the bacterial culture; (2) a protease-sensitive, dialyzable material from root extracts promoted attachment to root hairs but not to other epidermal cells; (3) free azospirilla cells incubated with seedlings for 2 to 4 days acquired an envelope surrounded by electron-dense granular material and azospirilla absorbed to root hairs after short-term incubation were also associated with granular material; and (4) addition of nitrate resulted in degranulation and inhibition of attachment. Umali-Garcia *et al.* (1980) considered these characteristics to be compatible with the lectin recognition model of the legume–*Rhizobium* association (Dazzo 1980; Dazzo and Hubbell 1975).

In legumes it is not clear to what extent lectins, as opposed to other factors such as polar bodies, are involved in nonspecific attachment of rhizobia (Schmidt 1979). The possibility that surface interactions of the types involved in rhizobia (Dazzo 1980) or pathogen recognition (Bradshaw-Rouse *et al.* 1981; Sequeira and Graham 1977) would allow differentiation of *Azospirillum* strains has not been reported.

Patriquin *et al.* (1983) have observed pronounced differences in nitrogenase activity by *Azospirillum* spp. associated with genetically different sugarcane calli, and these differences correspond to differences in attachment of *Azospirillum* to plant cells in suspension cultures. There were no pronounced differences between the two *Azospirillum* species in their attachment to individual host cells. Attachment to suspension cells was polar and associated with the presence of granular and fibrillar material.

Patriquin *et al.* (1983) found that attachment of a nonhomologous *A. brasilense* strain (Sp7), but not of a homologous strain (JM6B2 and

SP114), to maize root hairs was strongly inhibited by nitrate and to a lesser extent by ammonium. This suggests that interactions with nitrate could selectively affect the attachment of various strains from mixed rhizosphere populations.

Not yet understood is the significance of capsules or cysts in the attachment processes (see Section 2.1). The capsule illustrated by Dazzo (1980) surrounds a single *Rhizobium* cell, and extracellular polymers such as microfibrils in flocs have been reported to play a role in aggregating the infecting cells of *Rhizobium* (Napoli *et al.* 1975) and *Agrobacterium* (Matthysse *et al.* 1981) to the roots. The C-forms illustrated by Umali-Garcia *et al.* (1980) in *Azospirillum* surround two or more cells. The latter appears to be similar, if not identical, to the following: (1) the "C-forms" of *Azospirillum* associated with sugarcane callus, in which the capsule is proposed to confer some degree of oxygen protection (Berg *et al.* 1979, 1980); (2) the cystlike structures described by Lamm and Neyra (1981) which appear to confer resistance to desiccation; (3) cystlike forms considered to be involved in biphasic nitrogenase activity by *Azospirillum* in culture (Papen and Werner 1982); (4) sharply defined small colonies of apparently determinate size on surfaces of axenically grown wheat inoculated with *Azospirillum* (Patriquin *et al.* 1983) and barley with a microaerobic diazotrophic root isolate (Pohlman and McColl 1983); and (5) large cystlike structures observed in older, thick root segments of maize inoculated with *Azospirillum* (Ruscoe *et al.* 1978).

4.2 Deformation of root hairs

Some root hairs become deformed when legumes are grown with *Rhizobium*. Yao and Vincent (1969) found that the kind and degree of root hair deformations are correlated with the ability of different *Rhizobium* strains to nodulate the host. Branching and moderate curling of root hairs are affected by homologous or nonhomologous *Rhizobium* strains or their culture filtrates, both being more numerous in the presence of homologous strains.

Patriquin *et al.* (1983) noted branched root hairs in field-grown wheat seedlings and a greater frequency of these in seedlings inoculated with *Azospirillum brasilense* strain Sp245 than in uninoculated seedlings or in those inoculated with strains Sp7 and Sp107st. Seven *Azospirillum* strains induced more deformations of root hairs of wheat than did strains of *Rhizobium leguminosarum, Azotobacter chroococcum*, and *Escherichia coli* (Jain and Patriquin 1984). Wheat root hair branching was classified into two types: (1) branches of equal length ("tuning-fork deformation") and (2) branches of unequal length. Wheat seedlings, grown in semisolid agar in test tubes or germinated on water agar plates, exhibited the tuning-fork deformation only in the presence of *Azospirillum*. Unequally branched root hairs were observed in most uninoculated controls. The order of frequency of the tuning-form deformations in seedlings inoculated with different strains was consistently *A. brasilense* strain 245 > strain 107 > strain Sp7 > *Azospirillum* sp. strain 242. This order corresponds to the order of total N gains in wheat following inoculation of field-grown wheat with strains 245, 107, and 242 (Baldani *et al.* 1983). Strains Sp245 and Sp107st could be established within wheat roots while strain Sp7 could not (Baldani *et al.* 1986a). Strains 245 and Sp107st are homologous isolates from wheat roots while Sp7 and Sp242 are nonhomologous to wheat. Few or no tuning-fork deformations were produced in the presence of the nonhomologous strains. Jain and Patriquin (1984) observed that among several *Azospirillum* strains, Sp245 produced the highest concentrations of indole acetic acid (IAA) in culture and could reproduce the tuning-fork deformation in root hairs of wheat with IAA. Addition of N resulted, in most cases, in higher numbers of deformations; this was in contrast to reported inhibition of legume root hair deformation in the presence of ammonium or nitrate (Dart 1974). Strain 245 of *A. brasilense* inhibited nodule formation in clover and increased plant growth without nodule formation (Plazinski and Rolfe 1985c).

Root hair deformations were also observed in roots of field-grown maize, but Patriquin *et al.* (1983) were unable to induce such deformation in seedlings grown on water agar plates exposed to 10 different *Azospirillum* strains or in maize grown in semisolid agar with two homologous strains.

4.3 Root infection

The first strong experimental evidence for cereal root infection by diazotrophs came from experiments of Okon *et al.* (1977a) who found reduction of azospirilla numbers in inoculated field-grown maize roots from 7 × 10^6 cells/g root (washed roots) to 1.1 × 10^3 cells/g root after sterilization (2.5 minutes in sodium hypochlorite) and an increase to 6.5 × 10^6 cells/g root when these roots were crushed.

Azospirilla are commonly observed in intercellular spaces between the epidermis and the cortex, and also in the outer cortical layers of axenically grown grasses inoculated with *Azospirillum* (Patriquin and Döbereiner 1978; Umali-Garcia *et al.* 1978). Fluorescent antibody staining techniques have confirmed that such regions are colonized in field-grown plants (Schank *et al.* 1979). The demonstration of pectinolytic activity by *Azospirillum* in culture (Tien *et al.* 1981; Umali-Garcia *et al.* 1978) and the observation of clear zones following the contour of *Azospirillum* in the middle lamellae of inoculated guinea grass in monoxenic systems (Umali-Garcia *et al.* 1978) both suggest that penetration of these regions is an active process. Invasion of cortical tissues apparently remains intercellular as long as host cells are healthy.

Several workers have observed *Azospirillum* inside of collapsed root hairs and adjacent epidermal cells or inside rare cells in the cortex (Lakshmi *et al.* 1977; Patriquin and Döbereiner 1978; Umali-Garcia *et al.* 1980). Lakshmi *et al.* (1977) observed that when 15-day-old rice seedlings were dipped in suspensions of *Azospirillum*, colonization of the mucilaginous sheath surrounding root hairs was profuse within 30 minutes and the root hairs were soon filled with bacteria. The colonization started from the tip of the root hairs and gradually extended toward the base. In other plant species, penetration of root hairs and epidermal cells required up to 15 days. Invasion of epidermal cells via the root hairs may be analogous to instances in which rhizobia penetrate legume root hairs too rapidly for a controlled response by the host and the host cell dies (Sprent 1979). There is no evidence so far that azospirilla form viable intracellular associations with grasses.

On the basis of observations on the distribution of 2,3,5-triphenyltetrazolium chloride (TTC) reducing bacteria and the recovery of *Azospirillum* from roots of field-grown maize treated for up to 6 hours with 1% chloramine-T (a sterilizing agent), (Döbereiner and De-Polli 1981; Patriquin and Döbereiner 1978) it was concluded that azospirilla invade intercellular spaces in the inner cortex, stele, and lumen of xylem vessels and do so without disrupting the outermost cortex of the endodermis. The common occurrence and longitudinal distribution of bacteria in the cortex, and occasionally in the stele of branches, and a radiating pattern of colonization of the stele of main roots in the region of branches suggested that the initial colonization of these inner areas occurs in the regions of root branches and then spreads longitudinally, extending even into the stem of maize (Magalhães *et al.* 1979) and sugarcane (Patriquin *et al.* 1980; Rennie *et al.* 1982). Entry into roots could occur through disrupted cortical tissues where the branches emerge from the main roots. This region is believed to be the principal venue of infection of some legumes by rhizobia (Dart 1974) and one venue of invasion of monoxenically grown grasses by *Azospirillum* (Umali-Garcia *et al.* 1978). Similar infection venues were reported for nonfluorescent pseudomonads that can colonize extensive areas of xylem vessels of *Citrus* spp. (Gardner *et al.* 1985) and for enterobacteria that infect wheat (Lindberg *et al.* 1985). However, it is difficult to understand how the invasion of roots via these cortical disruptions could result in invasion of the stele without disrupting the endodermis. Magalhães *et al.* (1979) observed colonization of field-grown maize by

TTC-reducing bacteria within 2 cm of root apices. In cross sections just behind the root apices, tetrazolium-reducing bacteria were observed in the cortex and in the adjacent region of the stele. They suggested that initial entry into the stele occurs at apical regions where there is no appreciable thickening of the endodermal walls or of the Casparian strip or where the stele has not differentiated as such. This process may resemble that involving vascular nematodes whose larvae penetrate the root in the region of cell elongation prior to cell differentiation (Van Gundy and Freckman 1977). Some of the mechanisms that a plant normally uses to resist bacterial invasion of roots may not be operative at the root apex. It is of interest that initial events in the colonization of soybean (Bhuvaneswari *et al.* 1980) and alfalfa (Anollés and Favelukes 1986a, b) roots by *Rhizobium* take place at the root apex before there is visible differentiation of root hairs. Okon (1985) observed highest concentrations of *A. brasilense* on the root apex (Figure 4.1).

Localization of azospirilla in the inner regions of the root could be expected to facilitate efficient exchange of substrates and products between azospirilla and the host. Low p_{O2} (Crawford 1976) and the presence of malate (Butz and Long 1979) in the xylem might be especially conducive to N_2 fixation. However, with the exception of the report of Lakshmi *et al.* (1977), there have been no reports of colonization of these inner regions in axenically grown plants by azospirilla. A possible argument against the xylen colonization by azospirilla is that its pH, about 5.3 (Butz and Long 1979), would be unfavorable for N_2 fixation by *Azospirillum* (Day and Döbereiner 1976). Observations with several *Azospirillum* strains revealed that 2 out of 17 *A. brasilense* and 3 out of 8 *A. lipoferum* root isolates grew equally well or better at pH 5.7 than at pH 7.0, while none of the soil isolates grew at that pH (Döbereiner 1983).

There are no observations relating N_2 fixation or yield response to a particular site or region of colonization by *Azospirillum*. There are, however, a number of observations that are explained by assuming some type of internal localization or existence of protective structures:

1. The population of microorganisms recovered from surface-sterilized roots differs markedly from that recovered from untreated roots and is enriched in diazotrophs in general (Patriquin and McClung 1978) or by particular *Azospirillum* strains (Baldani and Döbereiner 1980).
2. Colonization via tips or points of emergence of lateral roots would explain various observations relating root activity to the presence of lateral roots (Döbereiner and Day 1976; Patriquin and McClung 1978) and the presence of an intact cortex (Döbereiner and Day 1976).

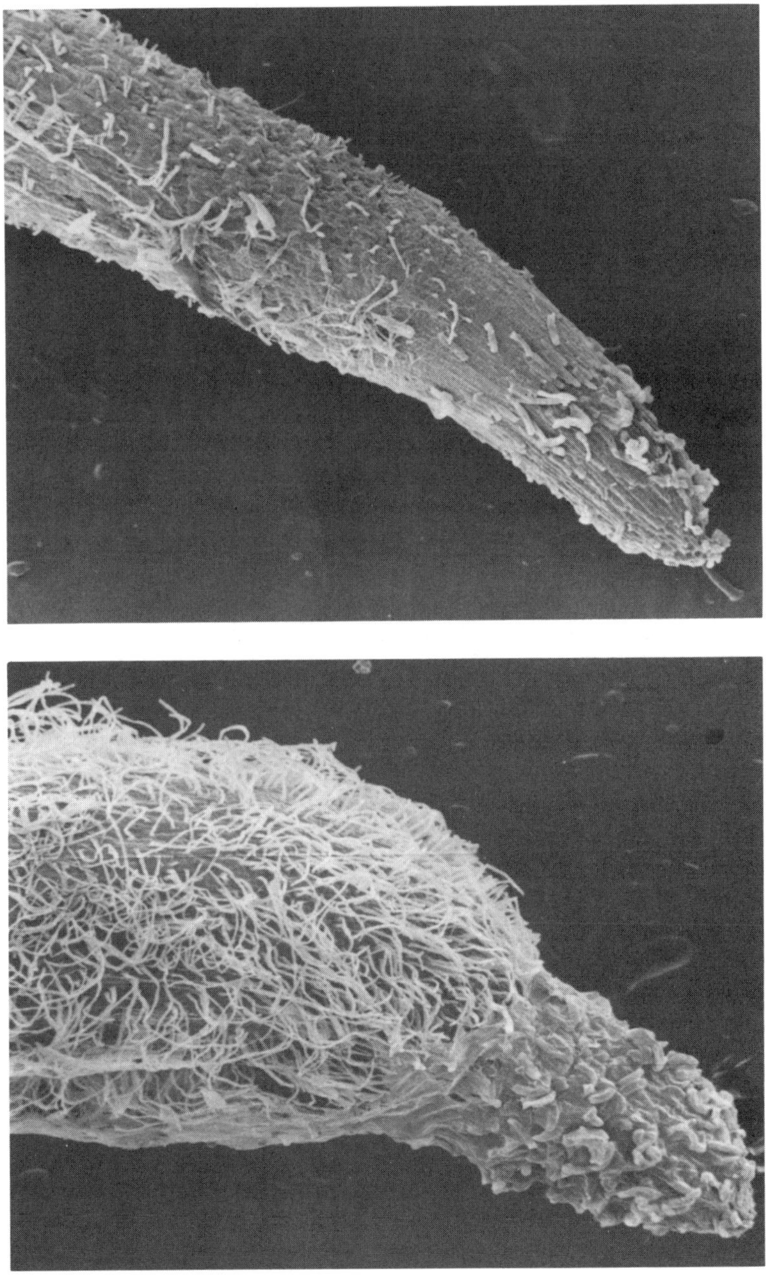

(a) (b)

Figure 4.1 Scanning electron micrograph showing the effect of inoculation with *A. brasilense* on root tip morphology of sorghum. (a) Control. (b) Inoculated (400×). (Reproduced from Okon 1985, by permission of Elsevier Science Publishers B. V., Amsterdam.)

3. N_2 fixation is low in chromosome substitution lines of wheat that have "leaky" roots and high in lines that have nonleaky roots (Rennie and Larson 1979).
4. Total N accumulation in wheat inoculated with various *Azospirillum* strains (Baldani *et al.* 1983) and nitrogenase activity of excised roots of various forage grasses (Souto and Döbereiner 1985) is correlated with numbers of azospirilla recovered from surface sterilized roots, but not with numbers recovered from untreated roots.

Azospirilla numbers are always higher on the root surface than within roots, and were constant on washed roots during the growth cycle of maize (Scott *et al.* 1978) and pearl millet (Rao and Venkateswarlu 1985), and increased in surface sterilized roots from almost zero in young plants to 10^5 or 10^6 during reproductive growth. This increase in *Azospirillum* numbers parallels increases in whole plant nitrogenase activity during flowering and grain filling. Root extracts from plants collected at this growth stage stimulated *Azospirillum* growth (Rao and Venkateswarlu 1985; Xavier *et al.* 1982), while extracts from roots collected in early growth phases inhibited *Azospirillum* growth.

4.4 Host plant affinities

Once infection of healthy root tissues by various diazotrophs was confirmed, it became evident that there must be differences between *Azospirillum* species and strains in their capacity to establish plant associations. Plant roots are always exposed to a multitude of soil microorganisms, although it is evident that each plant is susceptible to infection by only a few of them (Panopoulos *et al.* 1984). The special properties that enable microorganisms to penetrate plant tissues is a major research theme in plant pathology. Equally important is the study of the properties that make plants susceptible to infection (Panopoulas *et al.* 1984). The similarity of the situation of root-associated diazotrophs with that of plant pathogens suggests that analogies between the two systems may exist. Nevertheless, little information is available and nothing is known about the mechanisms that determine bacterial "virulence" and host plant "resistance" in association with diazotrophs.

The first and most striking example of host plant diazotroph specificity is that of *Paspalum notatum* with *Azotobacter paspali* (Döbereiner 1966), although infection of root tissues in this association has so far not been demonstrated. Nitrogenase activity on excised roots incubated under low p_{O2} has been attributed to *A. paspali* localized within the mucigel (Döbereiner *et al.* 1972).

The data in Table 3.3 show that *A. paspali* occurs almost exclusively on the root surface of one tetraploid ecotype of *Paspalum notatum* (cv. Batatais), a grass used for lawns because of its ability to spread rapidly in unfertile soils; 98% of 252 samples of this grass cultivar contained the bacteria. An improved cultivar of the same species used in Florida as forage grass (cv. Pensacola) and two other *Paspalum* spp. did not harbor the bacteria (Döbereiner 1970). Neither did several other Gramineae and many other plants. This association seems to be the most specific plant–bacteria association known to date. Neither in plant pathogens nor in the *Rhizobium* symbiosis is such absolute and apparently obligate specificity with one ecotype of one species known. Unfortunately, the nature of this association has not been further studied.

Much less defined are the interactions of *Azospirillum* spp. with cereals (Table 4.3). Certain host genotype–strain "specificities" have been observed between wheat cultivars (Millet *et al.* 1985). When maize, wheat, and rice were grown in pots inoculated with a mixture of various *Azospirillum* strains, infection of the inner root tissue also showed a certain degree of host–bacteria affinity (Baldani and Döbereiner 1980). In maize roots 58% of the isolates were *A. lipoferum* and within wheat roots 92% were *A. brasilense*. In a subsequent field experiment, regardless of heterologous inocula, surface-sterilized maize roots contained predominantly (81 to 90%) *A. lipoferum* while wheat roots contained predominantly (81 to 94%) *A. brasilense*. Furthermore, these data showed that the great majority of strains invading the roots were resistant to low levels (about 20 ppm) of streptomycin as had been observed with the total rhizosphere population (Table 2.12, page 60). Much lower numbers of streptomycin-resistant azospirilla were found in soil (Baldani and Döbereiner 1980).

Further data of Rocha *et al.* (1981) and Döbereiner and De-Polli (1980) indicated that roots of cereals and grasses possessing the C_4 photosynthetic pathway (e.g., maize, sorghum, *Panicum maximum, Pennisetum purpureum, Digitaria decumbens, Hemarthria altissima*, and *Cynodon dactylon*) are preferentially infected by *A. lipoferum*, whereas C_3 plant roots (e.g., wheat, rye, and oat) are primarily infected by *A. brasilense*. The discrimination in root infection among *Azospirillum* species and strains has been confirmed by serological reactions (De-Polli *et al.* 1980). Also, when strains were isolated from roots and marked with antibiotic resistance levels greater than those naturally occurring in soil or roots and then inoculated into field grown cereals, all strains seemed to establish in the soil but only certain homologous strains were able to dominate inner root tissue colonization (Table 4.3).

Table 4.3 Host–plant affinities in the infection of field-grown cereals

Inoculation strain	Origin	% inoculated strain[a] Soil	% inoculated strain[a] Surface sterilized roots	% increase of total plant N	Reference[b]
Wheat					
A. brasilense Sp245	Wheat roots	44	80	25	A
A. brasilense Sp107st	Wheat roots	50	80	15	A
A. brasilense Cd	Cynodon roots	53	11	−20	A
A. lipoferum Sp241st	Maize roots	—	—	7	B
Maize					
A. lipoferum Sp241st	Maize roots	—	—	37	B
A. brasilense Sp109st	Rice roots	—	—	20	B
Sorghum					
A. lipoferum S82	Sorghum roots	—	72	—	C

[a]Antibiotic-resistant marker strains were used as inoculants and the percentage of the inoculated strain within the *Azospirillum* population was evaluated as stated in the various papers cited.

[b]References are as follows: A, Baldani *et al.* 1986a; B, Freitas *et al.* 1982; C, Baldani *et al.* 1986a.

4.5 Growth substances

Azotobacter (Brown and Burlingham 1968), *Rhizobium* (Dart 1974), many pathogens (Dekhuijzen 1976), and various plant growth-promoting bacteria or other soil bacteria (Brown 1974; Krasil'nikov 1958) produce plant hormones. In pure culture, *Azotobacter paspali* (Barea and Brown 1974) and *Azospirillum brasilense* (Tien *et al.* 1979) produce gibberelin and cytokinin-like substances and also auxin from tryptophan. Inoculation of nutrient solutions around roots of pearl millet with *A. brasilense* caused increases in plant dry weight and in numbers of root hairs and lateral roots, but no change in root weight and decreases in the length of the main root (Tien *et al.* 1979). These effects could be mimicked by supplying a combination of indoleacetic acid, gibberelin GA-3, and kinetin. Others (Ferreira *et al.* 1986; Kapulnik *et al.* 1981b; Martin and Glatzle 1982) have observed increases in root masses following inoculation. Umali-Garcia *et al.* (1980) found that the effects of inoculation on root hair morphology and lateral root morphogenesis were suppressed by nitrate [5 m*M* $Ca(NO_3)_2$].

Hormonal interactions might also be responsible for certain changes in shoot morphology associated with inoculation, for example, increased diameter of maize stems (Kapulnik *et al.* 1982) and increased tillering and number of ears in wheat (Ferreira *et al.* 1986; Reynders and Vlassak 1982). The possibility of obtaining hormone over-producing mutants of *Azospirillum* (Hartmann *et al.* 1983b; Kolb and Martin 1985) will help to answer this question.

Okon (1982) cites experiments in which root segments from inoculated plants exhibited 30 to 50% greater uptake of NO_3^-, P_2O_5, and K^+ than did root segments from uninoculated plants. Jain and Patriquin (1984) reported enhanced P and K assimilation by plants inoculated with *A. brasilense* strain Sp245 while strain Sp7 reduced P and K incorporation. No gross changes in root morphology were noted, but cross sections of roots revealed altered, more irregular arrangements of cells in the outer four to five layers of the cortex (Okon 1982). This suggests that *Azospirillum* "softens" the middle lamellae through the action of pectinolytic enzymes (Tien *et al.* 1981; Umali-Garcia *et al.* 1978), thus enhancing the mineral absorption surface of cortex cells in a kind of "sponge" effect (Okon 1982). Penetration of the cortex can take place without collapse of cells such as occurs when pathogenic bacteria producing large quantities of pectinases invade roots (Tien *et al.* 1981). Plant growth substances (IAA, gibberellic acid, or kinetin in concentrations of up to 1 ppm) had no effect on *Azospirillum* growth rates or respiration rates nor any effect on cell morphology (Silveira-Schrank and Drozdowicz, 1983).

4.6 Tissue culture associations

The simplification of the plant component achieved by plant cell or tissue culture is appropriate for studying certain aspects of plant–diazotroph associations. The observation that cowpea rhizobia derepressed nitrogenase activity when grown on surfaces, in intercellular spaces, or in close proximity to callus of various nonleguminous plants (Child and Kurz 1978; Scowcroft and Gibson 1975) demonstrated that neither nodulation nor intracellular infection (nor even legume hosts) were essential for nitrogenase activity by these rhizobia. Child and Kurz (1978) established associations of *A. brasilense* Sp7 with tissue cultures of several legumes and nonlegumes for periods of up to 2 weeks. The callus nutrient solution was not nitrogen-free but some nitrogenase activity was recorded. This activity was higher in associations with nonlegumes than with legumes and was always higher than that observed in similar tissue cultures inoculated with *Rhizobium* spp.

Associations of *A. brasilense* Sp7 with tissue cultures of sugarcane were established and survived continued growth with subculturing for periods as long as 18 months even in the absence of added fixed N (Vasil *et al.* 1979). Similar tissue cultures of tobacco, pearl millet (*Pennisetum americanum*), and centipede grass (*Eremochloa ophiuroides*) were also established for shorter periods with *A. brasilense*, but the bacteria inhibited callus growth. Inoculation of tobacco or sugarcane cells with *Azotobacter vinelandii* caused a similar inhibition of callus growth. Electron microscope studies (Berg *et al.* 1979) revealed that *A. brasilense* never entered healthy sugarcane cells, but grew profusely in association with the callus tissue, with as many as 2×10^8 bacteria/mm^2 found on the surface of the callus. An interesting observation was the presence in these callus cultures of large, round, encapsulated forms of *Azospirillum* (C-forms) usually containing several cells and considerable quantities of poly-β-hydroxybutyrate (Figure 2.3, page 7). Berg *et al.* (1980) observed that the C-forms were more common when the organisms were growing under N-fixing conditions and unfavorably high p_{O_2}, and they suggested that the lipopolysaccharide capsule gave oxygen protection for the nitrogenase. Functional associations (up to 200 nmol C_2H_4/root/hour) with *Azospirillum* were also obtained with excised roots of maize, sorghum, and millet (O'Hara *et al.* 1983).

4.7 Interaction with nitrate reduction

The capacity for assimilatory and dissimilatory NO_3^- reduction in *Azospirillum* spp. was discussed in Section 2.3. There are several ways in which plant–bacteria associations interact with NO_3^- assimilation. The capacity for dissimilatory NO_2^- reduction (nir^+), which is the key step in denitrification, seems to be more common in soil isolates (58, 56, and 30% of rhizosphere isolates from wheat, rice, and maize being nir^+, respectively) than in root isolates (Baldani and Döbereiner 1980). Strains that do not denitrify (nir^- strains) predominate within roots (100% and 96% for wheat and rice, respectively) indicating competition advantages for strains that dissimilate NO_3^- to NO_2^-, but not further (Baldani and Döbereiner 1980; Döbereiner and Baldani 1981). This is corroborated by the observation that nitrate reductase negative mutants of *A. brasilense* Sp245, a strain that establishes itself within wheat roots (Baldani *et al.* 1986a), do not compete well with other soil microorganisms in colonizing the wheat histosphere. When wheat plants were inoculated with a nir^+ *A. brasilense* strain (Sp7), denitrification was demonstrated without the need for additional carbon sources (Bothe *et al.* 1983; Neuer *et al.* 1985) and nif^- mutants of the same strain (Pedrosa and Yates 1984) showed three to four times higher denitrification rates than the parent strains (Neuer *et al.* 1985). The latter authors also showed that *A. lipoferum* strain Sp59 denitrified in association with wheat.

The nitrate reductase of azospirilla also interferes in the plant–bacteria association. It is responsible for the inhibition of nitrogenase activity by NO_3^- and nitrate reductase negative mutants (NR^-) can fix N_2 in culture and in plants in the presence of high levels of NO_3^- (Ferreira *et al.* 1987; Magalhães *et al.* 1978). On the other hand, in experiments carried out in large (60-cm-diameter) concrete cylinders containing soil amended with $K^{15}NO_3$, inoculation with two *A. brasilense* nir^- strains Sp245 and Sp107st enhanced fertilizer assimilation causing enrichment of the ^{15}N atom percent excess relative to the uninoculated controls, instead of the dilution that would have been expected if atmospheric $^{14}N_2$ had been fixed and incorporated (Table 4.4) (Boddey *et al.* 1986). In a field experiment with the same strains, a nitrate reductase negative mutant of strain Sp245 did not establish within roots but only in the rhizosphere soil (as strain Sp7), and unlike the parent strain (Sp245) it failed to enhance N incorporation. In monoxenic wheat cultures in test tubes, inoculation with three NR^- mutants of this strain caused much higher increases in nitrate reductase activities in leaves than did inoculation with the parent strain, indicating that the bacterial nitrate reductase contributes to NO_3^- reduction in the roots (Ferreira *et al.* 1986).

The elucidation of the interaction of the nitrate reductases of plants and bacteria with N_2 fixation in cereal associations is of fundamental importance for a better understanding of the systems, and represents

Table 4.4 Effect on N incorporation of inoculation of wheat with various *A. brasilense* strains[a]

Inoculant strain	Total N in grain (g/cylinder)[b]	Atom % ^{15}N excess	^{15}N recovery (mg ^{15}N excess/cylinder)
Control	0.866	0.156	1.33
Sp245[c]	1.271	0.171	2.20
Sp107st[c]	1.195	0.190	2.30
Sp7[d]	1.276	0.159	2.00

[a]From Boddey *et al.* 1986.

[b]Concrete cylinders (60 × 50 cm) filled with fumigated soil fertilized with $K^{15}NO_3$.

[c]Isolated from surface sterilized wheat roots.

[d]Isolated from rhizosphere soil of *Digitaria decumbens*.

the basis for a better use of these associations. The associations must accommodate biological N_2 fixation with NO_3^- assimilation from fertilizer complements because it appears improbable that associative N_2 fixation alone will be able to supply all necessary N for high grain yields in cereals.

5

Quantification of N_2 Fixation

The quantification of associative N_2 fixation in cereals and grasses is more complicated than that in legumes. While well-nodulated legumes grown in sterile N-free sand or vermiculite pots can fix all the N they need for optimal growth, grasses can not. Although the standard method for ***Rhizobium*** selection involves the use of sterile Leonard jars, attempts to quantify associative N_2 fixation by such systems are faced with the following problems:

1. N_2-fixing bacteria inoculated in the absence of the normal soil microflora may be unable to reduce the p_{O_2} to levels low enough to permit N_2 fixation.
2. In cereals, and probably also in other grasses, the peak of maximum N_2 fixation occurs between flowering and grain filling, and it is difficult to grow cereals to that late growth stage in Leonard jars or other sterile growth vessels.
3. It seems impossible to maintain azospirilla-free Leonard jars by the standard methods used for ***Rhizobium*** selection. This we attribute to the fact that soils in tropical regions usually contain 10^5 to 10^7 cells of azospirilla/g of soil while rhizobia frequencies are on the order of 10^2 to 10^3/g soil. Unavoidable dust carries azospirilla and therefore contamination is 100 to 1000 times easier than by rhizobia.

Problems encountered in the field or in pots with soil where plants can be grown to maturity include the following:

1. In most tropical soils N_2-fixing bacteria, especially azospirilla, occur in high numbers, and thus the addition of inoculant strains may not cause their establishment, or even if it does, the established strain might not be more effective than the autochthonous strains.
2. Production of growth substances (see Section 4.5) and enhancement of NO_3^- assimilation can result in larger plants with an equal or higher percentage of N and overlap with N_2 fixation effects.
3. The high background total N, even in N-deficient soils, makes N quantification difficult.

All these difficulties make the sample comparison of an uninoculated (nonfixing) plant with an inoculated one (supposedly fixing) very inconvenient for the quantification of N_2 fixation in Gramineae. The methods that have been giving most reproducible results will therefore be discussed in some detail.

5.1 Balance studies

Nitrogen balance studies seem the most logical approach for measuring N_2 fixation in the field. The amount of nitrogen removed from the soil is most directly estimated from soil nitrogen contents at the beginning and at the end of the experiment subtracted from the total N content of harvested plants. Such studies, however, have to be conducted for long periods and with many replicates since small losses in soil N are very difficult to estimate with sufficient precision against the very large total N content in the soil. In temperate agricultural systems the best documented case of N gains associated with grasses or cereals is the Broadbalk experiment conducted at the Rothamsted Experimental Station in England beginning in 1843 until the present. Here, where only wheat has been grown for over 140 years without any N fertilizer addition, a positive balance of 25 kg N/ha/year was observed and was attributed mainly to algal N_2 fixation (Day *et al.* 1975). Parker (1957a) followed N changes under grass and legume cover during 3 years and found positive N balances of 50 kg N/ha under grass cover and 80 kg N/ha under legumes. Less known are long-term N balances in India (Dart 1986) where after a sequence of 44 crops of grains, grasses, cotton, and sugarcane, the soil N content increased more than the amount of fertilizer added. In an 8-year period of Guinea grass (*Panicum maximum*) growth in Nigeria, a positive balance of 57 kg N/ha/year was observed (Dart 1986). App *et al.* (1984) found highly positive balances in long-term rice plantations in the Philippines. In two localities after

24 and 17 rice crops, N gains of 79 and 103 kg N/ha/year, respectively, could be attributed to associative N_2 fixation, while algae N_2 fixation was estimated as 6 to 8 kg N only. No data of specific plant–bacteria associations in long-term field balance studies are available.

More precise estimates of balance studies are obtained in pots where there is no possibility of N from the subsoil being translocated to the soil surface and where the soil can be completely homogenized. Several earlier experiments (Giddens 1977; Moore 1977; Purchase 1978) gave strong evidence for considerable N enrichments in pots with soil planted with grasses, but the first direct proof of N_2 fixation in a cereal came with the experiments carried out in the Philippines (App *et al.* 1980). In these experiments care was taken to eliminate all possible accretions of N from the atmosphere and also that from cyanobacteria (cover with a black cloth) to measure solely associative N_2 fixation with rice. To obtain significant differences the soil had to be carefully sieved and mixed before planting rice which was grown for four to six consecutive crops. Triplicate Kjeldahl analyses were performed for plant samples and three composite soil samples were analyzed from each pot in duplicate Kjeldahl analyses. By using all these precautions it was possible to obtain significantly positive balances in three experiments where the percentage of N_2 fixed in relation to the total N assimilated represented at least 14 to 18%.

Loureiro (1985) studied in more detail the methodology for precise N balance studies. A predigestion with Devarda's alloy according to Liao (1981) increased the recovery of NO_3^- in soil from 82 to 96%. Also, plant material predigestion with Devarda's alloy was necessary. The variability was reduced when the soil samples were ground and sieved through 100 mesh sieves, and at least 2 g of soil had to be analyzed for better reproducibility. Soil sampling turned out to be the largest source of variation. The methods used resulted in standard deviations of less than 1.5% for means of three analyses of the same sample.

Another problem in pot balance studies is the loss of N by leaching and/or denitrification which cannot be quantified by correcting with unplanted control pots, as losses there were found to be much higher (App *et al.* 1980; Loureiro 1985). In one experiment with four *Brachiaria* spp. these losses were corrected according to the amount of ^{15}N-labeled fertilizer recovered; positive balances between 100 and 268 mg N/pot were obtained after five cuts, representing 30 to 70% of the total N accumulation (Loureiro 1985). However, after summing up the various errors in soil analyses, soil sampling, plant analyses, and fertilizer losses, the N balances were not statistically significant.

Several experiments in pots with vermiculite and sand mixtures showed substantial N gains for *Paspalum notatum* transplanted from the field into vermiculite (110 mg N/pot for 60 days) (Döbereiner and Day 1975).

Likewise, sorghum planted in vermiculite inoculated with root macerates (124 to 269 mg N/pot for 49 days) (Dart 1986) or maize and setaria inoculated with *Azospirillum* spp. (Cohen *et al.* 1980; Kapulnik *et al.* 1981a) gave positive responses. These experiments probably gave statistically significant results because the error source in the N analyses was smaller than for experiments in soil, although the problem of summing up errors from the various analyses remains.

5.2 $^{15}N_2$ incorporation and ^{15}N isotope dilution

The $^{15}N_2$ incorporation method directly measures the incorporation of molecular N_2 into the soil plant system, while the ^{15}N isotope dilution method evaluates N_2 fixation indirectly by estimating the dilution of ^{15}N-labeled fertilizer or soil N by $^{14}N_2$ derived from the atmosphere. Clearly the evaluation of $^{15}N_2$ incorporation is the most direct and reliable method for the evaluation of N_2 fixation. The few data already obtained with this method give clear evidence of N_2 fixation and incorporation into plant tissue in forage grasses (De-Polli *et al.* 1977; Morris *et al.* 1985), rice (Eskew *et al.* 1981; Ito *et al.* 1980), and sorghum (Giller *et al.* 1984). The data are summarized in Table 5.1. Giller *et al.* 1984 described a specially devised chamber for $^{15}N_2$ fixation assays which he used for millet and sorghum. The amounts of N taken up by the plants have been small, but in most cases the plants were exposed at a younger growth stage than that at which maximal rates of nitrogenase activity (C_2H_2) have been observed (Von Bülow and Döbereiner 1975; Watanabe *et al.* 1978). The major restriction to this method is financial. $^{15}N_2$ gas is very expensive and therefore experiments have to be performed with small plants closed up in restricted gas volumes. This is best achieved by exposing the plants to $^{15}N_2$ during a 24- to 72-hour period and analyzing them after additional growth periods. For high precision it is desirable that highly labeled gas ($>$50 atom % $^{15}N_2$) be used in such experiments. In the apparatus developed by De-Polli *et al.* (1977), soil–plant cores taken from the field in steel cylinders of 9 cm diameter were transferred to cleaned oil cans with the same diameter. Active cores were selected by C_2H_2 reduction assays (see below) and placed into closed, gas-tight vessels provided with a gas sampling port. The gas phase was first replaced with helium or argon and then adjusted with $^{15}N_2$ (70 to 90 atom % enrichment) to a p_{N_2} of 0.5 atm. Adequate light and temperature controls are essential, and CO_2 and O_2, which are monitored by gas chromatography, should be adjusted whenever necessary. More precise results can be obtained with gas lysimeter assemblies, such as the apparatus proposed by Witty and Day (1978).

Table 5.1 Evaluation of N_2 fixation by $^{15}N_2$ incorporation

Plant species	Plant growth conditions	Incubation		Atom % ^{15}N excess	μg N_2 fixed/core/day	Reference
		Hours	Atom % $^{15}N_2$			
Digitaria decumbens	10-cm cores of field-grown plants	72	85.5			De-Polli *et al.* 1977
Roots				0.151		
Rhizomes				0.146	9.8	
Leaves				0.007		
Paspalum notatum	10-cm cores of field-grown plants	72	85.5			De-Polli *et al.* 1977
Roots				0.563		
Rhizomes				0.703	17.6	
Leaves				0.070		
Oryza sativa	Pots	312	60			Eskew *et al.* 1981
Roots				0.898		
Shoot				0.141	56.2	
Panicle				0.344		
Oryza sativa	17-cm cores	168	72			Ito *et al.* 1980
Roots				0.714		
Leaf sheath				1.570	45.6	
Leaves				0.003		
Sorghum vulgaris	Small pots	72	16			Giller *et al.* 1984
Roots				0.040		
Tops				0.067	15.3	
Crab grass	10-cm cores	24	10			Morris *et al.* 1985
Roots				0.0035		
Tops				0.0051	64.4	
Barn yard grass	10-cm cores	24	10			Morris *et al.* 1985
Roots				0.0017		
Tops				0.0047	11.8	
Bermuda grass	10-cm cores	24	10			Morris *et al.* 1985
Roots				0.0022		
Tops				0.0029	33.3	

The ^{15}N isotope dilution technique (McAuliffe *et al.* 1958) is based on the use of ^{15}N-labeled soil or fertilizer which is "diluted" in plants that assimilate $^{14}N_2$ from the atmosphere in symbiosis or associations with microorganisms and where the percentage of this dilution is proportional to the percentage of N_2 fixed in a system. These estimates seem straightforward in sand cultures supplied with fertilizer containing known amounts of atom percent ^{15}N excess. But once soil is used, ^{14}N liberated from the soil organic matter is also assimilated by the plant and deep-rooting plants assimilate more ^{14}N from lower soil horizons than more shallow-rooting plants (Boddey and Victoria 1986). In addition, if fertilizer is applied at the beginning of the experiment, fast-growing plants assimilate more ^{15}N at the early growth stage when the fertilizer concentrations are still high.

For all these reasons "nonfixing" control plants have been used with similar growth habits to the test plants (Boddey and Victoria 1986; Witty 1983). The control plants and test plants should have similar recovery of the fertilizer N as estimated by the total ^{15}N assimilation, indicating that comparable soil volumes have been explored. For the quantification of associative N_2 fixation, which is usually less than 20% of the total N incorporated, concrete cylinders filled with soil uniformly mixed with ^{15}N-labeled organic matter (plant material from previous ^{15}N experiments can be used) has been found to be most useful (Boddey *et al.* 1984; Boddy and Victoria 1986). Because of the constant slow release of ^{15}N, assimilation by control plants is easier to match with assimilation by test plants (Witty 1983). For estimates of N_2 fixation in the grass *Paspalum notatum* cv. Batatais the nonfixing ecotype Pensacola was used as control, and both ecotypes recovered almost equal total amounts of ^{15}N. Estimates from these experiments indicated that 10% of the plant N had been derived from the atmosphere amounting to 20kg N/ha (Boddey *et al.* 1983). The available data obtained with the ^{15}N isotope dilution method are summarized in Table 5.2. In the sugarcane experiment shown in Table 5.3, a cultivar that did not show a positive balance was used as the nonfixing control. Because very highly negative balances were obtained in the pots without plants (Lima *et al.* 1986) and because some N also was lost in the planted pots, the cultivar considered to be nonfixing must have obtained some N from the atmosphere and, therefore, the value for sugarcane in Table 5.3 is probably an underestimate.

The data discussed above were all obtained without artificial inoculation with N_2-fixing bacteria. The comparison of inoculated plants with sterile or poorly infected plants is another way of quantifying the contribution of biological nitrogen fixation (BNF). In wheat, inoculation with *A. brasilense* strain Sp245 in concrete cylinders with soil mixed with ^{15}N-labeled fertilizer did not produce isotope dilution but rather enrichment, as discussed in Section 4.6 (Table 4.4). Christiansen-Wen-

Table 5.2 Quantification of N_2 fixation by the ^{15}N isotope dilution method

Plant species	Plant growth conditions	Non-N_2-fixing control	Atom % $^{15}N_2$ excess: Control	Atom % $^{15}N_2$ excess: Test plant	Estimate of N_2 fixation: % of total N incorporation	Estimate of N_2 fixation: Time to fix 1 kg/ha	References
Paspalum notatum	Concrete cylinder, repeated ^{15}N fertilizer applications	*P. notatum* cv. Pensacola	0.563	0.506	9.9	21 yr	Boddey *et al.* 1983
Brachiaria decumbens	^{15}N-labeled organic matter	*B. radicans*	0.248	0.180	29.6	29 yr	Boddey *et al.* 1986
B. decumbens	^{15}N-labeled organic matter	*B. radicans*	0.248	0.152	39.1	45 yr	Boddey *et al.* 1986
Panicum maximum ecotype K190	^{15}N-labeled organic matter	*B. radicans*	0.299	0.185	38.7	2.9 months[a]	Miranda and Boddey 1986
P. maximum cv. colonião	^{15}N-labeled organic matter	*B. radicans*	0.0299	0.021	29.3	2.2 months[a]	Miranda and Boddey 1986
Sacarum officialis cv. CB 47–89	64-liter pots with mineral ^{15}N applied	*S. officialis* cv. IAC 59–150	0.1097	0.059	47.0	~120 yr	Lima *et al.* 1986
Poa pratense	Pots inoculated with *Klebsiella pneumoniae*	Uninoculated control	0.640	0.593	7.3	—	Haatela and Kari 1986
Leptochloa fusca	Pots inoculated with *A. brasilense*	Uninoculated control	9.840	4.96	49.6	—	Malik and Zafar 1985

[a]Summer months only.

Table 5.3 Differences between sugarcane cultivars in N_2 fixation[a]

Cultivar	Total plant N accumulated in 2 years (g/65-liter pot)	Total N balance[b] (g/65-liter pot)	Atom % ^{15}N excess
CB 47–89	34.8	+25.30	0.0586
CB 47–355	16.5	+ 8.90	0.1015
IAC 52–150	13.5	+ 3.20	0.1097
NA 56–79	13.0	+ 1.99	0.1047
Unplanted	0	− 9.68	—

[a]From Lima *et al.* 1986.
[b]Initial soil N + fertilizer N + total plant N − final soil N.

iger *et al.* (1985) also found low BNF inputs in sorghum inoculated with various *A. lipoferum* strains (2 to 2.5%); most of the fixed N remained in the roots (25%). Malik and Zafar (1985), however, growing Kallar grass (*Leptochloa fusca*, a desert grass) in monoxenic cultures in test tubes with gravel, observed much higher proportions of BNF-derived N in plants inoculated with *A. brasilense* (Table 5.2).

In pots with sand and watered with ^{15}N-containing nutrient solution (Haahtela and Kari 1986) the grass *Poa pratense* obtained as much as 34% of its shoot N from biological fixation when a *Klebsiella pneumoniae* strain isolated from roots of the same species was inoculated. Wheat inoculated with the same strain derived only 16% of the N from fixation, and total dry matter and N incorporation was not increased significantly by inoculation.

5.3 The acetylene reduction method

The acetylene (C_2H_2) reduction test has been widely used since its description by Schöllhorn and Burris (1966) and Dilworth (1966). It is based on the observation that C_2H_2 is reduced to C_2H_4 by all nitrogenase enzymes known to date; C_2H_4 can be easily measured with high sensitivity by flame ionization gas chromatography. Therefore, it is the rate of C_2H_2 reduction activity of the nitrogenase that is measured and not its rate of N_2 reduction ($N_2 \rightarrow NH_3$). Detailed discussions on the principles of the C_2H_2 reduction method have been exhaustively reviewed (Eskew and Ting 1977; Witty 1979). This sensitive method can therefore be useful for qualitative evaluations of N_2 fixation by certain bacteria or plant–bacteria associations and their screening. It also has been widely used for comparisons of different bacteria, treatments, or systems. But it does not give reliable results on its own for the quantification of N_2 fixation, although it can be useful for the confirmation of results with one or more of the previously described methods.

Uprooted nodulated legumes lose up to 50% of their nitrogenase activity after 5 to 10 minutes exposure to C_2H_2 (Minchin *et al.* 1983), and nonnodulated Gramineae roots maintain practically no or very low activity, especially when incubated under air. After prolonged preincubation under low p_{O2}, the activities are recovered. A probable explanation for this is the drastic change in p_{O2}, and therefore for excised root C_2H_2 reduction assays it is best to place the entire root systems immediately into large assay vessels with water under a N_2 stream. The water is then replaced by N_2 and the reaction is started by the injection of 1 to 2% O_2 and 10% C_2H_2 (Souto and Döbereiner 1984). Many other variations of this method have been proposed (e.g., De-Polli *et al.* 1982; Van Berkum and Bohlool 1980), some of which have shown results in good agreement with intact core assays. Comparisons of excised root assays with intact soil–plant core assays taken in the field should be made for every specific environment and system. In forage grasses and small grains, statistically significant correlations of these two assay methods were obtained (Boddey *et al.* 1978; Döbereiner 1978b; Nery *et al.* 1977), and comparable results were also reported by Weier and Date (personal communication), Witty (1979), Patriquin and Denike (1978), and Souto and Döbereiner (1984), in which excised roots usually underestimate the intact core rates.

Difficulties in removing large cores without seriously disturbing the system make such comparisons difficult if large root systems (e.g., maize, sorghum, or sugarcane) are to be assayed (Tjepkema and Van Berkum 1977; Wani *et al.* 1983). Slow gas diffusion rates are another problem in assays with large cores (Patriquin and Denike 1978; Witty 1979).

The excised root assay has been widely criticized (Barber *et al.* 1976; Eskew and Ting 1977; Koch 1977; Tjepkema and Van Berkum 1977) because multiplication of diazotrophic bacteria has been observed during the 8- to 16-hour preincubation period, which seems to invalidate the method. The above-mentioned correlations with intact core assays using small grasses and grains, however, indicate at least the validity of relative estimates done by this method. In addition, no nitrogenase activity can be found in roots of young maize plants (Pereira *et al.* 1978; Von Bülow and Döbereiner 1975) even after prolonged preincubation periods, although the numbers of diazotrophs present are as high as during the reproductive growth phase (10^7 or more) (Magalhães *et al.* 1979; Rao and Venkateswarlu 1985), and roots excised from very active grass cores show zero nitrogenase activity during the preincubation period. These observations indicate that the lag before the onset of C_2H_2 reduction in the excised root assay is more complex and can not be explained by multiplication of bacteria alone. For additional discussion on the lag in C_2H_2 reduction on excised roots, see Döbereiner (1978b).

There are many variables among the C_2H_2 reduction method, but the intact soil–plant system seems to be the most reproducible. Several methods have been proposed (Balandreau 1975a; Dart *et al.* 1972; DePolli *et al.* 1977; Lee *et al.* 1977a; Witty and Day 1978) in which the natural soil structure is maintained, and therefore microsites with specific p_{O2} regulation systems should be expected to remain intact. Such cores show linear C_2H_2 reduction rates during short incubation periods (Balandreau 1975). Over longer periods, however, these rates are no longer linear because of the pronounced day–night cycle in grasses (Balandreau *et al.* 1974; Döbereiner and Day 1975). The day–night variations in nitrogenase activity are caused by photosynthate availability, but also by changes in insolation, temperature, and soil and air humidity. Attempts to obtain quantitative estimates must therefore include measurements during at least 24 hours.

There are large numbers of estimates of N_2 fixation in grasses and cereals by intact core C_2H_2 reduction assays and these vary from very low (e.g., Tjepkema and Van Berkum 1977) to very high estimates (e.g., 150 to 750 nmol C_2H_4/g roots, Döbereiner and Day 1975; 30 to 620 nmol, Balandreau *et al.* 1975). In few cases, however, have there been comparisons with N balance studies or ^{15}N methods.

The C_2H_2 reduction method is also being widely used to measure nitrogenase activity in bacterial cultures. Good agreement of integrated C_2H_2 reduction rates over the whole growth period of a culture with the total N_2 fixed (conversion factor C_2H_2:N_2 = 3 to 4) were obtained when low concentrations of carbon substrates were used (Day and Döbereiner 1976), and it is easy to compare various diazotrophic species or strains. But there is no evidence so far that nitrogenase activity of cultures has any significant correlation with the performance of the organism in the field or that such performance can be used for the selection of strains for field inoculation. These evaluations, when performed during the log phase of growth, measure the potential growth rates on N_2 of a certain strain, but give no information on the capacity of this strain to compete and grow in the rhizosphere or to colonize roots. Hence they give little information on the capacity of this strain to fix N_2 in root associations.

More informative are C_2H_2 reduction rates by intact plants in monoxenic test tube cultures; these tests measure nitrogenase activities relying on energy sources supplied by living plants and therefore give estimates of the performance of plant–bacteria associations of seedling stages. Such assays have been used for selection of bacterial strains and plant genotypes (Bouton *et al.* 1985; Dommergues and Rinaudo 1979; Ela *et al.* 1982; Jagnow 1985), although in only two cases have the results been confirmed by measurements on adult potted or field plants (Charyulu *et al.* 1985; Wani *et al.* 1985). Many other data have also shown that inoculation of plants grown in pots with soil with various diazotrophs can enhance nitrogenase activity (C_2H_2) (Christiansen-Weniger *et al.* 1985); Cohen *et al.* 1980; Nur *et al.* 1980a), but proper strain selections for cereal or grass inoculation have so far not been done.

6

Agronomic Implications

There are many ways in which associations of N_2-fixing bacteria with crop plants can be exploited for the benefit of agriculture. The most immediate way seems to be the inoculation of field plants with N_2-fixing bacteria, but in many cases plant selection or plant breeding directed to enhance such associations or the identification of environmental factors, soil conditions, and fertilizer management systems that favor N_2 fixation can yield more promising results.

6.1 Strain selection and field inoculation

Inoculation of plants with beneficial bacteria on the farm, with a few exceptions, is carried out only for legumes. The legume symbiosis has been known for a century and has been studied and explored extensively all over the world. The procedures recommended for general use are as follows: (1) selection of superior *Rhizobium* strains in terms of (a) high capacity to nodulate and outcompete indigenous populations, (b) high efficiency of N_2 fixation, (c) efficient translocation of fixed N to the seeds, and (d) tolerance to adverse environmental conditions; (2) establishment of the selected strains in the rhizosphere; (3) selection and breeding of legume host genotypes for symbiotic performance; and (4) soil fertilization and management to maximize N_2 fixation.

These points give an example of the complexity of problems to be solved before reliable and reproducible field responses to inoculation of cereals with N_2-fixing bacteria can be expected. In fact, even in legume symbiosis statistically significant responses to inoculation with *Rhi-*

zobium in the field occur only in less than half of the conducted experiments.

It is indeed surprising that there are so many reports of azospirilla inoculation success (e.g., Avivi and Feldman 1982; Bouton *et al.* 1979; Desale and Konde 1984; Döbereiner and Baldani 1982; Freitas *et al.* 1982; Kapulnik *et al.* 1981b; Millet *et al.* 1985; Venkateswarlu and Rao 1983), especially since most of them used strain Sp7 or Cd, a red derivative of it (Eskew *et al.* 1977), which proved to be much weaker root colonizers than others (see Section 4.1). The probable reason for the success of strains Sp7 and Cd, especially in Israeli soils and in soils of temperate climates, is that azospirilla are absent or occur only sporadically in these soils, and therefore even strain Sp7 effectively contributes to the development of the associated plant. In soils where high numbers of azospirilla are normal, as in most tropical soils, inoculation of wheat with strain Sp7 + Cd resulted in no significant effect (Boddey *et al.* 1986), or even in a significant decreases of percent N and total N content in grains (Baldani *et al.* 1987).

These observations together with the results on host plant affinities (see Section 4.3) leave no doubt that extensive strain selections have to precede field inoculation programs. No satisfactory method for strain selection, however, has been developed yet. The difficulties in comparison with *Rhizobium* strain selection have been discussed in Section 4.1, and so far field experiments seem to be the most promising procedure (Venkateswarlu and Rao 1983). The recent progress in establishing selected and labeled (antibiotic-resistant) *Azospirillum* strains within roots that coincide with field responses (Baldani *et al.* 1986) may suggest easier and more efficient methods for strain selection, at least for wheat. There are two strains (Sp245 and Sp107st), both isolated from surface sterilized wheat roots, which in various field experiments with wheat were shown to comprise more than 80% of the inoculated azospirilla populations within roots (Baldani *et al.* 1986). They increased plant weights in monoxenic test tube culture (Bazzicalupo *et al.* 1985; Ferreira *et al.* 1987) and induced the most reproducible field responses among many other strains tested. No such superior strains have so far been found for other cereals. Interaction of azospirilla with other unidentified diazotrophs and wheat cultivars were observed by Millet *et al.* (1985). In many other experiments the effects of inoculation have been extremely variable, from nonsignificant, negative, or zero effects on yield parameters, to large, highly significant increases in dry matter and total crop nitrogen content. It should be stressed that there have been several field and pot experiments at the author's institute where inoculation even with homologous strains has not resulted in significant dry matter or plant N increases. This may also be true for other laboratories, as insignificant effects tend to remain unpublished.

While large percent yield increases with inoculation may appear to be important, they can be misleading if they occur only in very low yielding, highly nitrogen-deficient plants. Smaller inoculation effects may represent larger N inputs in crops with moderate to high yields. The former is true of the data of Rennie (1980b). Although he found that 12.6% of the N in 8-week-old inoculated maize plants came from N_2 fixation, these plants had a less than 0.5 g dry weight and had a nitrogen content of only 0.55%. Similarly, inoculation of maize plants in sterile N-free culture by Nur *et al.* (1980a) resulted in a 44% increase in dry matter, but the 5-week-old plants had only a dry weight of 0.85 g. However, other experiments of Nur *et al.* (1980a) and other authors (Baldani *et al.* 1986b; Baltensperger *et al.* 1978; Cohen *et al.* 1980; Kapulnik *et al.* 1981b; Smith *et al.* 1976) have all shown significant dry matter and/or N yield increases in either field or pot experiments with plants of reasonable size.

Smith *et al.* (1976) found that field inoculation of *Panicum maximum* was most successful when the plants were receiving moderate levels of N fertilizer (20 to 60 kg N/ha). This was also evident from the results of Taylor (1979) and Kapulnik *et al.* (1981b), the latter using Jubilee maize. Similar results have led some workers (Tien *et al.* 1979) to suggest that in view of the possible production of plant hormones by *Azospirillum* spp., yield increases due to inoculation could have been caused by hormonal effects on plants. Evidence for this also comes from the frequent observation that estimates based on acetylene reduction activity fail to account for the increase in N content in the inoculated plants (Bouton *et al.* 1979; Bouton and Zuberer 1979).

The possible role of bacterial nitrate reductase (NR) in nitrate assimilation by the plant has been discussed in Section 4.7. The most reproducible and highest responses so far recorded were obtained with strain Sp245, which could be established within roots and promoted increases not only of plant growth but also especially of N incorporation into grains (Baldani *et al.* 1986) (Table 2.3). This seems to be due at least in part, to a more efficient NO_3^- assimilation with Sp245. Nitrate reductase negative mutants of this strain did establish on the root surface but not within roots (Table 6.1) and had no effect on total N incorporation (Boddey *et al.* 1986; Ferreira *et al.* 1987).

An interesting and very different approach is the use of azospirilla inoculants in connection with *Rhizobium* which leads to improved legume growth. Plazinski and Rolfe (1985a) found that the inoculation of clover with mixtures of rhizobia and azospirilla at the ratio 1:500 increased nodule numbers, but certain *Azospirillum* strains made nodules ineffective and in some cases higher ratios inhibited nodulation. Strain Sp245 cited above as the most effective wheat strain was also most active in these experiments on clover. This strain, when applied in the

Table 6.1 Effect of inoculation with various strains of *Azospirillum* spp. on field-grown wheat in southern Brazil[a]

Inoculant strain[d]	% inoculated strain[b] RS	RI	Plant tops (g/plot)[c]	Total N in tops (g/plot)[c]	Grain yield (g/plot)
State of Rio de Janeiro					
Control	0	0	32.4b[e]	0.66c	23.2c
Sp7	50	0	44.0b	0.87c	27.0c
Sp246	94	27	61.4a	1.29b	35.1b
Sp245	100	67	71.1a	1.57a	41.5a
Sp245 NR−	81	0	42.4b	0.82c	25.6c
A. amazonense	56	11	60.5a	1.26b	31.9b
State of Paraná					
Control	—		278	6.37b	27.8
Sp245	—		365	9.00a	36.5
Sp245 NR−	—		338	5.44b	33.8
A. amazonense	—		326	7.60ab	21.6

[a]From Baldani *et al.* 1986a, b, and Boddey *et al.* 1986.

[b]% of isolates of the inoculated strain identified by its antibiotic-resistance marker. RS, root surface; RI, root interior.

[c]At flowering stage in Rio de Janeiro and at grain-filling stage in Paraná.

[d]*A. brasilense* strains Sp7 (ATCC 29245) isolated from rhizosphere soil of Digit-grass; Sp246 from washed and Sp245 from surface sterilized wheat roots; *A. amazonense* strain YTr isolated from washed wheat roots.

[e]Numbers within each column followed by the same letter are not statistically different from each other.

ratio of 1:2000, completely inhibited nodulation but enhanced plant growth (Plazinski and Rolfe 1985c). Synergistic inoculation effects of rhizobia and azospirilla were reported under field conditions with soybeans (Singh and Subba Rao 1979), soybeans and winged beans (Iruthayathas *et al.* 1983), cowpeas (Singh *et al.* 1980), and clovers (Plazinski and Rolfe 1985b). Pronounced synergistic effects of azospirilla with mycorrhiza in barley suggest another promising approach (Subba Rao *et al.* 1985a, b).

6.2 Establishment of inoculated strains in the field

Because diazotrophs, especially azospirilla, are so widely distributed in soil, inoculation practices can only have success if strains superior to those already present in the soil can out-compete the native diazotrophs and the general soil microflora and become established in the rhizosphere and within roots. This situation can be illustrated by attempts to inoculate promiscuous forage legumes with "cowpea" group rhizobia which resulted in only 5% of the nodules being formed by the inoculated strain (Date 1971). It is therefore of crucial importance that inoculated azospirilla were recently found to establish themselves promptly in soils and to represent the majority of the azospirilla population on or within roots of wheat or sorghum (Baldani *et al.* 1986). These strains had been labeled by selection of spontaneous double antibiotic resistant mutants and were identified in the highest dilutions in MPN counts. The data in Table 6.1 give an example for wheat. So far strains isolated from surface-sterilized roots establish themselves within roots (Baldani *et al.* 1986), whereas strains isolated from washed roots establish themselves only on the root surface (Table 6.1).

Similar results were obtained from experiments with field-grown sorghum in which 72% of the isolates from surface sterilized roots were identified as the inoculated *A. lipoferum* strain (SpS82), which had been isolated originally from surface sterilized sorghum roots (Baldani *et al.* 1986). In these experiments intact individual root pieces were capped with paraffin wax, surface sterilized, and then preincubated in azospirilla medium. Only root pieces showing no growth were crushed and used for further isolation of bacteria. This method guarantees that only organisms localized inside the roots were recovered.

Other attempts to establish inoculated *Azospirillum* strains (Albrecht *et al.* 1983; Smith *et al.* 1984) or other bacteria (Wright and Weaver 1981) in field-grown plants failed. In these experiments, however, nonhomologous diazotrophs were used as inoculants, in most cases strain Sp7 (Albrecht *et al.* 1983), supplied in extremely high numbers.

The identification of certain *Azospirillum* strains by immunoagglutination, as has been used for rhizobia strain identification, is difficult because serogroups contain many strains, although clear differentiation between species (*A. lipoferum* versus *A. brasilense*) is possible (Dazzo and Milam 1975; De-Polli *et al.* 1980; Ladha *et al.* 1982; Schank *et al.* 1979). The distinct serogroups for a number of *A. lipoferum* isolates from C_4 plant roots (De-Polli *et al.* 1980) which were confirmed by immunodiffusion tests (M. J. Sampaio, personal communication) unfortunately have not been studied further.

6.3 Plant genotype effects

As in legumes, the plant genotype plays an important role in the effectiveness of associative N_2 fixation. Plant selection and breeding for cereal or forage grass genotypes that stimulate N_2-fixing bacteria are promising prospects for enhancing diazotroph multiplication and N_2 fixation. This becomes evident from the pronounced differences between plant genotypes of the same genus or species. Examples discussed in Section 4.3 are the differences in root colonization by diazotrophs between *Paspalum notatum* ecotypes (Döbereiner 1970), and those between certain chromosome substitution lines in wheat (Larson and Neal 1978). Differences in C_2H_2 reduction activities of intact soil plant systems of paddy rice that could be correlated to root weights ($r = 0.605$, $p = 0.0001$) (Lee *et al.* 1977b) suggest that breeding for large root systems might be an approach.

Similar field core screenings evaluated by the C_2H_2 reduction test showed differences between genotypes of pearl millet (Bouton and Brooks 1982; Bouton *et al.* 1985; Dart and Wani 1982; Manga *et al.* 1985), sorghum (Dart and Wani 1982), and wheat (Avivi and Feldman 1982; Nery *et al.* 1977). Attempts to develop seedling tests by the C_2H_2 reduction method, which would make breeding programs much easier (Kipe-Nolt *et al.* 1985), showed large differences between genotypes of rice (Dommergues and Rinaudo 1979; Rinaudo 1977), maize (Ela *et al.* 1982), and other cereals (Jagnow 1985), but are only valid if the results are confirmed by ^{15}N methods or balance studies. Charyulu *et al.* (1985) showed correlations of seedling C_2H_2 reduction tests with evaluations of natural ^{15}N abundance in field-grown rice.

Screening for qualitative or quantitative differences in root exudates (Balandreau *et al.* 1981; Kipe-Nolt *et al.* 1985) or organic acid contents in roots (Stumpf *et al.* 1977; S. Urquiaga-Caballero, personall communication) is another approach. Differences in malic and transaconitic acid concentrations in roots of sorghum were suggested to be responsible for differences in tolerance to Al^{3+} toxicity (Cambraia *et al.* 1983), and one Al^{3+} resistant sorghum cultivar showed higher intact C_2H_2 reduction activity during the growth cycle than a sensitive cultivar (Christiansen-Weniger *et al.* 1985). Manga *et al.* (1985) found that both dominant and recessive genes of pearl millet governed associative nitrogenase activity. He observed significant correlations with small ears and early flowering and suggested these characteristics for field selection.

Recent results obtained with the ^{15}N isotope dilution method indicate large differences between *Panicum maximum* ecotypes (24 to 38% derived from BNF) (Miranda and Boddey 1986), and sugarcane cultivars (see Table 5.3). These results were confirmed by N balances in 60-liter pots (Lima *et al.* 1986).

6.4 Environmental factors

High soil moisture has been shown to be positively correlated with C_2H_2 reduction activity (ARA) in grasses in Canada, England, the United States, and Australia (Day *et al.* 1976; Tjepkema and Burris 1976; Vlassak *et al.* 1973; Weier *et al.* 1981). The ARA values of roots from dryland rice were lower than those of wetland rice (Yoshida and Ancajas 1970). In maize and sorghum soil moisture was also shown to be positively correlated with ARA (Balandreau *et al.* 1978; Hegazi *et al.* 1983; Wani *et al.* 1983). It is likely that roots and rhizosphere have more microaerophilic sites in which diazotrophs can be active under high moisture levels; this is consistent with the observation that low p_{O2} encourages maximal nitrogenase activity in several crops (Döbereiner *et al.* 1972; Patriquin 1978; Souto and Döbereiner 1984; Trolldenier 1977). High moisture may also facilitate nutrient exchange between soil and plant roots, and thus improve the overall physiology of both partners. The optimal temperature for azospirilla growth and N_2 fixation is 36°C and very poor growth occurred below 27°C (Neves *et al.* 1976). Excised roots of maize and rice showed similar temperature limits for ARA (Neves *et al.* 1976; Van Berkum and Sloger 1982). Similar results were reported for maize from *in situ* ARA assays, the activity increasing from 18 to 35°C (Van Berkum and Sloger 1982).

Diurnal fluctuations in ARA associated with intact cereals and grasses have been recorded (Balandreau *et al.* 1974; Döbereiner and Day 1975; Nery *et al.* 1977; Van Berkum and Day 1980; Watanabe and Lee 1977), which suggest that there is a close link between photosynthesis, translocation, and N_2 fixation. At constant temperature, however, very little day–night differences in ARA values were reported for rice (Watanabe and Cabrera 1979), but ARA of intact rice plants decreased (Van Berkum and Sloger 1982) within 15 minutes after transferring the plants into the dark even though the root temperature was maintained at 35°C. *Spartina alterniflora* incubated continuously in the dark exhibited ARA similar to plants maintained in the normal day–night cycle for at least 13 days, suggesting that a large carbon reserve was present in the root system; the presence of this carbon reserve pool was demonstrated with $^{14}CO_2$ (Boyle and Patriquin 1981). The size of this pool was dependent on the rate of photosynthesis which in turn depended on light intensity, temperature, and other factors controlling carbon assimilation.

Soil fertility is another environmental factor affecting nitrogenase activity. The addition of 10 μg/ml of urea nitrogen dramatically reduced the ARA of excised roots (Trolldenier 1977) as well as of intact plants (Watanabe and Cabrera 1979). This inhibition was seen on root surfaces or in rhizosphere soil but not in surface sterilized roots (Boyle and Patriquin 1980). Inhibition of ARA on excised sorghum roots reached 50% with 1.4 μg/ml ammonium or nitrate nitrogen, and the level of combined nitrogen required for complete inhibition was 140 μg/ml (Neyra and Van Berkum 1976). Thus, it is apparent that if plants are provided with sufficient nitrogen, they will not develop associated ARA.

Significant negative correlations have been observed between the percentage of nitrogen in root and associated ARA, indicating that the availability of nitrogen in soil regulates nitrogenase activity (Van Berkum and Sloger 1981). Low fertilizer nitrogen applications (20 kg/ha of NH_4NO_3) did not influence ARA in field grown *Pennisetum purpureum* and *Digitaria decumbens* (Döbereiner and Day 1975) whereas heavy applications of nitrogen fertilizers inhibited ARA (Pereira *et al.* 1981) indicating that a reasonable starter dose of fertilizer nitrogen is not harmful to nitrogenase activities. The application of NO_3^- fertilizer to field grown maize plants resulted in inhibition of azospirilla growth in cultures amended with root extracts of these plants (Xavier *et al.* 1982). Root extracts from nonfertilized plots also inhibited azospirilla growth at early growth stages of the plant, but enhanced it at the flowering stage of maize. These effects coincide with the pronounced changes of nitrogenase activity that occur, with a peak at flowering and grain filling observed in several cereals (Kapulnik *et al.* 1981b; Van Berkum and Sloger 1982; Von Bülow and Döbereiner 1975; Watanabe *et al.* 1978).

Effects of other soil fertility factors apparently have not been studied. Certainly phosphate fertilization and organic matter should be expected to enhance N_2 fixation. Application of Mo to *Brachiaria decumbens* pastures under grazing increased dry matter production and total N accumulation during the whole growth cycle by 19 and 29%, respectively. During this period it also significantly increased azospirilla numbers within roots but not in rhizosphere soil (Miranda *et al.* 1985).

7

Potentials and Prospects for the Future

There seems to be no *a priori* reason preventing diazotrophs from associating with crop plants and from forming successful partnerships capable of contributing significantly to the N requirement of the crop (Patriquin *et al.* 1983). As Parker (1957b) pointed out, there is no particular reason why the association should be enclosed in a specific visible structure such as a nodule. Once it has been established that a viable nitrogen fixer can infect, multiply, and retain its ability to fix N_2 within the root or rhizosphere, the contribution of this fixation to the nitrogen nutrition of the plant will depend on the following factors:

1. The quantity of carbon substrate in the root or rhizosphere available to the microflora present;
2. the proportion of this substrate that is used by the N_2-fixing organisms;
3. the efficiency with which these microorganisms utilize the substrate to fix N_2; and
4. the availability of this fixed N to the plant.

All plants contribute organic substrates to the rhizosphere (Rovira 1969). Most estimates of the quantity of such substrates have been made in sterile solution culture. However, it has been demonstrated (Barber and Gunn 1974; Boulter *et al.* 1966) that exudation is greatly enhanced when the roots are in contact with a solid medium and that non-water-soluble material is quantitatively more important than the soluble fraction (Martin 1975). It has also been demonstrated (Barber and Lynch

1977; Barber and Martin 1976) that the presence of microorganisms in the rhizosphere greatly increase the quantity of carbon substrate released, and *Azospirillum* inoculation has been shown to enhance root exudation and root permeability (Venkateswarlu and Rao 1985). Barber and Martin (1976) and Martin (1977) estimated that between 14 and 25% of the carbon assimilated by nonsterile wheat and barley plants was released into the soil.

Bacterial counts can give an idea of the relative amounts of carbon substrates present in the rhizosphere that are available to the N_2-fixing organisms. However, as has been pointed out before (Döbereiner and De-Polli 1980), the great variation in size and specific metabolic activity of different bacterial species indicates that the presence of equal numbers of different bacteria at a particular site does not mean that they are utilizing equal quantities of the available substrates.

The efficiency with which free-living aerobic bacteria utilize carbon substrates to fix atmospheric nitrogen has been reported by many authors (e.g., Hill 1978; Jensen 1965). Maximal efficiencies measured in traditional batch cultures are on the order of 25 mg N_2 fixed/g C source consumed. However, under conditions of low p_{O_2} (Hill 1976; Postgate 1971) and low carbon concentrations (Becking 1971; Day and Döbereiner 1976; O'Toole and Knowles 1973), the carbon cost of N_2 fixation is considerably lower. Balandreau and Knowles (1978) suggested that these conditions could exist in the rhizosphere and that the efficiency of N_2 fixation in this zone could be in excess of 100 mg N_2 fixed/g C consumed. Volpon *et al.* (1981) and Stephan *et al.* (1981) have reported efficiencies of up to 200 mg N_2/g C (83 mg N_2/g lactate) and 120 mg N_2/g C (48 mg N_2/g glucose) in late log-phase cultures of *A. brasilense* and *A. lipoferum*, respectively, where O_2 and C were growth limiting.

Availability of fixed N to the plant is variable and depends on the performance of the various diazotroph species and strains in association with different plant species and thus can not be generalized. The regulatory system of associated diazotrophs would have to be modified, making it similar to that of rhizobia, in such a way that it effectively excretes fixed N. Although many data have shown increased N assimilation by plants inoculated with certain diazotrophs (see Section 6.1), there are only a few data that give unequivocal proof that fixed N was directly (or indirectly) transferred to the plant and translocated to the tops. Most results obtained with the isotope dilution method (Table 5.2) show incorporation of biologically fixed N over periods of several months or years and give little or no information on the associated diazotrophs nor on the process through which the fixed N was transferred to the plant. In contrast, the data on inoculation of Kallar grass with *Azospirillum brasilense* (Malik and Zafar 1985) and of *Poa pra-*

tense with *Klebsiella pneumoniae* (Haahtela and Kari 1986) in pot cultures (Table 5.2) furnish strong evidence that the N_2 fixed by a certain bacterium is made available to the plant within one growth period. Other similar experiments with sorghum inoculated with *A. lipoferum* (Christiansen-Weniger *et al.* 1985) indicated that, although 26% of the N accumulated in the roots originated from the atmosphere, very little (4%) was transferred to the tops, indicating that the fixed N remained within the bacterial cells in or on the roots.

The data obtained with $^{15}N_2$ incorporation experiments (Table 5.1) give more detailed information on the transfer of fixed N within short time periods. There is good evidence that $^{15}N_2$ is transferred to plant tops within 24 to 72 hours in forage grasses (De-Polli *et al.* 1977; Morris *et al.* 1985). In rice and sorghum incubated under $^{15}N_2$ for slightly longer periods there seemed to be little transference to the leaves in one experiment (Ito *et al.* 1980), but considerable amounts to the leaf sheaths of rice and to sorghum plant tops and rice panicles in two other experiments (Giller *et al.* 1984; Eskew *et al.* 1981). No $^{15}N_2$ incorporation experiments are available where inoculation with specific diazotrophs enhanced N incorporation into plants.

There are good prospects that in the near future substantial progress will be made in elucidating the mechanism of at least some of the diazotrophic associations. In the last decade six or seven new diazotrophs have been described which associate and infect roots (Baldani *et al.* 1986; Barraquio *et al.* 1983; Haahtela *et al.* 1983a; Magalhães *et al.* 1983; Seldin *et al.* 1984; Tarrand *et al.* 1978). Inoculation with certain strains of most of them has been shown to increase plant growth and N incorporation, and labeled strains could be specifically established within roots in the field. Large differences between plant genotypes in stimulating N_2 fixation and incorporation have permitted studies on the qualitative and quantitative occurrence of various diazotrophs, and the ^{15}N isotope methods have been improved and adapted for quantification of associative N_2 fixation. These developments represent important scientific bases for future identification of the mechanisms involved in the establishment of diazotrophic associations. Attempts to genetically manipulate plants and bacteria must be based on a detailed understanding of the physiology of each individual partner and of the association. Only by understanding what nature has achieved in millions of years will we be able to further improve upon nature for the benefit of mankind.

References

Abd-el-Malek, Y. 1971. Free-living nitrogen-fixing bacteria in Egyptian soils and their possible contribution to soil fertility. *Plant Soil, Spec. Vol.* 423–442.

Adams, M. W. W., Mortenson, L. E., and Chen, J. S. 1981. Hydrogenase. *Biochim. Biophys. Acta* 594:105–176.

Albrecht, S. L., Gaskins, M. H., Milam, J. R., Schank, S. C., and Smith, R. L. 1983. Ecological factors affecting survival and activity of *Azospirillum* in the rhizosphere. *In* Klingmüller, W. (ed.), Azospirillum *II: Genetics, Physiology, Ecology*. Birkhäuser, Basel, pp. 138–148 (Experientia supplementum, 48).

Allison, F. E. 1947. *Azotobacter* inoculation of crops. I. Historical. *Soil Sci.* 64:413–429.

Alvarez-Morales, R. A., and Lemos-Pastrana, A. 1980a. Quimiotaxis de *Azospirillum lipoferum* y *Azospirillum brasilensis.* II. Acción de los aminoácidos contenidos en exudados radiculares de gramíneas. *Rev. Lat.-Amer. Microbiol.* 22:137–142.

Alvarez-Morales, R. A., and Lemos-Pastrana, A. 1980b. Quimiotaxis de *Azospirillum lipoferum* y *Azospirillum brasilensis* hacia exudados radiculares de gramíneas. I. Actividad de maíz trigo y sorgo. *Rev. Lat.-Amer. Microbiol.* 22:131–135.

Anderson, K., and Shanmugam, K. T. 1977. Energetics of biological nitrogen fixation: determination of the ratio of formation of H_2 to NH_4^+ catalysed by nitrogenase of *Klebsiella pneumoniae in vivo. J. Gen. Microbiol.* 103:107–122.

Annollés, G. C., and Favelukes, G. 1986a. Quantitation of adsorption of rhizobia in low numbers to small legume roots. *Appl. Environ. Microbiol.* 58:371–376.

Annollés, G. C., Favelukes, G. 1986b. Host-symbiont specificity expressed during early adsorption of *Rhizobium meliloti* to the root surface of alfalfa. *Appl. Environ. Microbiol.* 52:377–382.

App, A., Watanabe, I., Alexander, M., Ventura, W., Daez, C., Santiago, T., and De Datta, S. K. 1980. Non-symbiotic nitrogen fixation associated with the rice plant in flooded soils. *Soil Sci.* 130:283–289.

App, A., Santiago, T., Menguito, C., Ventura, W., Tirol, A., Po, J., Watanabe, I., De Datta, S. K., and Roger, P. 1984. Estimation of the nitrogen balance for irrigated rice and the contribution of phototrophic nitrogen fixation. *Field Crops Res.* 9:17–27.

Avivi, Y., and Feldman, M. 1982. The response of wheat to bacteria of the genus *Azospirillum. Israel J. Bot.* 31:237–245.

Balandreau, J. 1975. Mesure de l'activité nitrogénasique des microorganismes fixateurs libres d'azote de la rhizosphere de quelques graminées. *Revue Écol. Biol. Sol* 12:273–290.

Balandreau, J. 1986. Ecological factors and adaptive processes in N_2-fixing bacteria populations of the plant environment. *Plant Soil* 90:73–92.

Balandreau, J., and Knowles, R. 1978. The rhizosphere. *In* Dommergues Y. R., and Krupa, S. V. (eds.), *Interactions Between Nonpathogenic Soil Microorganisms and Plants.* Elsevier Scientific, Amsterdam, pp. 243–268 (Developments in agricultural and managed-forest ecology, 4).

Balandreau, J., Millier, C. R., and Dommergues, Y. R. 1974. Diurnal variations of nitrogenase activity in the field. *Appl. Environ. Microbiol.* 27:662–665.

Balandreau, J., Rinaudo, G., Fares-Hamad, I., and Dommergues, Y. 1975. Nitrogen fixation in the rhizosphere of rice plants. *In* Stewart W. D. P. (ed.), *Nitrogen Fixation by Free-Living Microorganisms.* Cambridge University, Cambridge, pp. 57–70.

Balandreau, J., Ducerf, P., Hamad-Fares, I., Weinhard, P., Rinaudo, G., Millier, C., and Dommergues, Y. 1978. Limiting factors in grass nitrogen fixation. *In* Döbereiner, J., Burris, R. H., Hollaender, A., Franco, A. A., Nevra, C. A., and Scott, D. B. (eds.), *Limitations and Potentials for Biological Nitrogen Fixation in the Tropics.* Plenum, New York, pp. 275–302 (Basic Life Sciences, 10).

Balandreau, J., Richez, M., Weinhard, P., Tran, P. Q. U., and Ancelin F. 1981. Utilisation des cultures de racines en double alimentation pour l'etude de la fixation rhizosphèrique de l'azote. *Agronomie* 1:161–166.

Baldani, J. I. 1984. Ocorrência e caracterizacõ de *Azospirillum amazonense* em comparação com as outras espċies dêste gênero, em raízes de milho, sorgo e arroz. Itaguaí-RJ, UFRRJ. 110 p. M. S. thesis.

Baldani, J. I., Baldani, V. L. D., Xavier, D. F., Boddey, R. M., and Döbereiner, J. 1982. Efeito da calagem no número de actinomicetos e na % de bactērias resistentes a estreptomicina na rizosfera de milho, trigo e feijão. *Rev. Microbiol.* 13:250–263.

Baldani, J. I., Baldani, V. L. D., Sampaio, M. J. A. M., and Döbereiner, J. 1984. A fourth *Azospirillum* species from cereal roots. *An. Acad. Brasil Ciênc.* 56:365.

Baldani, J. I., Baldani, V. L. D., Seldin, L., and Döbereiner, J. 1986c. Characterization of *Herbaspirillum seropedicae* gen. nov., sp. nov., a root-associated nitrogen-fixing bacterium. *Int. J. Syst. Bacteriol.* 36:86–93.

Baldani, V. L. D., and Döbereiner, J. 1980. Host-plant specificity in the infection of cereals with *Azospirillum* spp. *Soil Biol. Biochem.* 12:433–439.

Baldani, V. L. D., Baldani, J. I., and Döbereiner, J. 1983. Effects of *Azospirillum* inoculation on root infection and nitrogen incorporation in wheat. *Can. J. Microbiol.* 29:924–929.

Baldani, V. L. D., Alvarez, M. A. B., Baldani, J. I., and Döbereiner, J. 1986. Establishment of inoculated *Azospirillum* spp. in the rhizosphere and in roots of field grown wheat and sorghum. *Plant Soil* 90:35–45.

Baldani, V. L. D., Baldani, J. I., and Döbereiner, J. 1987. Inoculation of field grown wheat with *Azospirillum* spp. in Brazil. *Biol. Fert. Soils* (in press).

Bally, R., Thomas-Bauzon, D., Heulin, Th., Balandreau, J., Richard, C., and De Ley, J. 1983. Determination of the most frequent N_2-fixing bacteria in a rice rhizosphere. *Can. J. Microbiol.* 29:881–887.

Baltensperger, A. A., Schank, S. C., Smith, R. L., Litell, R. C., Bouton, J. H., and Dudeck, A. E. 1978. Effect of inoculation with *Azospirillum* and *Azotobacter* on turf-type Bermuda genotypes. *Crop Sci.* 18:1043–1045.

Bani, D., Barberio, C., Bazzicalupo, M., Favilli, F., Gallori, E., and Polsinelli, M. 1980. Isolation and characterization of glutamate synthase mutants of *Azospirillum brasilense. J. Gen. Microbiol.* 119:239–244.

Barak, R., Nur, I., Okon, Y., and Henis, Y. 1982. Tactic responses of *Azospirillum brasilense* towards oxygen and organic compounds. *Israel J. Bot.* 31:229–236.

Barak,R., Nur, I., and Okon, Y. 1983. Detection of chemotaxis in *Azospirillum brasilense. J. Appl. Bact.* 54:399–403.

Barber, D. A., and Gunn, K. B. 1974. The effect of mechanical forces on the exudation of organic substances by the roots of cereal plants grown under sterile conditions. *New Phytol.* 73:39–45.

Barber, D. A., and Lynch, J. M. 1977. Microbial growth in the rhizosphere. *Soil Biol. Biochem.* 9:305–308.

Barber, D. A., and Martin, J. K. 1976. The release of organic substances by cereal roots into soil. *New Phytol.* 16:69–80.

Barber, L. E., and Evans, H. J. 1976. Characterization of a nitrogen-fixing bacterial strain from the roots of *Digitaria sanguinalis. Can. J. Microbiol.* 22:254–260.

Barber, L. E., Tjepkema, J. D., Russell, S. A., and Evans, H. J. 1976. Acetylene reduction (nitrogen fixation) associated with corn inoculated with *Spirillum. Appl. Environ. Microbiol.* 32:108–113.

Barberio, C., Bazzicalupo, M., Gallori, E., Polsinelli, M. 1982. Regulation of ammonium assimilation and N_2 fixation in *Azospirillum brasilense. In* Klingmüller, W. (ed.), Azospirillum: *Genetics, Physiology, Ecology*. Birkhauser, Basel, pp. 52–58. (Experientia supplementum, 42).

Barea, J. M., and Brown, M. E. 1974. Effects on plant growth produced by *Azotobacter paspali* related to synthesis of plant growth regulating substances. *J. Appl. Bact.* 37:583–593.

Barea, J. M., Ocampo, J. A., Azcón, R., Olivares, J., and Montoya, E. 1978. Effects of ecological factors on the establishment of *Azotobacter* in the rhizosphere. *In* Granhall, U. (ed.), *Environmental Role in Nitrogen-Fixing Blue-Green Algae and Asymbiotic Bacteria.* Swedish Natural Science Research Council, Stockholm, pp. 325–330 (Ecological Bulletins, 26).

Barraquio, W. L., Watanabe, I. 1981. Occurrence of aerobic nitrogen fixing bacteria in wetland and dryland plants. *Soil Sci. Plant Nutr.* 27:121–125.

Barraquio, W. L., De Guzman, M. R., Barrion, M., and Watanabe, I. 1982. Population of aerobic heterotrophic nitrogen-fixing bacteria associated with wetland and dryland rice. *Appl. Environ. Microbiol.* 43:124–128.

Barraquio, W. L., Ladha, J. K., and Watanabe, I. 1983. Isolation and identification of N_2-fixing *Pseudomonas* associated with wetland rice. *Can. J. Microbiol.* 29:867–873.

Bazzicalupo, M., and Gallori, E. 1983. Genetic analysis in *Azospirillum. In* Klingmüller, W. (ed.), Azospirillum *II: Genetics, Physiology, Ecology*. Birkhäuser, Basel, pp. 24–28 (Experientia supplementum, 48).

Bazzicalupo, M., Cresta, E., and Favilli, F. 1985. An "in vitro" assay for evaluating the *Azospirillum* wheat association. *In* Klingmüller, W. (ed.), Azospirillum *III: Genetics, Physiology, Ecology*. Springer-Verlag, Berlin, pp. 139–146.

Becking, J. H. 1959. Nitrogen fixing bacteria of the genus *Beijerinckia* in South African soils. *Plant Soil* 11:193–206.

Becking, J. H. 1961a. Studies on N fixing bacteria of the genus *Beijerinckia*. I. Geographical and ecological distribution in soils. *Plant Soil* 14:49–82.

Becking, J. H. 1961b. Studies on N-fixing bacteria of the genus *Beijerinckia*. II. Mineral nutrition and resistance to high levels of certain elements in relation to soil type. *Plant Soil* 14:297–323.

Becking, J. H. 1963. Fixation of molecular nitrogen by an aerobic vibrio or spirillum. *J. Microbiol. Serol.* 29:326.

Becking, J. H. 1971. Biological nitrogen fixation and its economic significance. *In: Nitrogen-15 in Soil-Plant Studies*. International Atomic Energy Agency, Vienna, pp. 189–222 (Panel proceedings series).

Becking, J. H. 1985. Pleomorphism in *Azospirillum. In* Klingmüller, W. (ed.), Azospirillum *III: Genetics, Physiology, Ecology*. Springer-Verlag, Berlin, pp. 243–262.

Beijerinck, M. W. 1925. Uber ein Spirillum, welches freien Stickstoff binden kann? *Centralbl. Bakt. II Abt* 63:353–357.

Berg, R. H., Vasil, V., and Vasil, I. K. 1979. The biology of *Azospirillum* sugarcane association. II. Ultrastructure. *Protoplasma* 101:143–163.

Berg, R. H., Tyler, M. E., Novick, N. J., Vasil, V., and Vasil, I. K. 1980. Biology of *Azospirillum* sugarcane association. I. Enhancement of nitrogenase activity. *Appl. Environ. Microbiol.* 39:642–649.

Bergersen, F. J. 1974. The effects of phosphate concentration on N_2 and NH_4^+ grown *Klebsiella pneumoniae. J. Gen. Microbiol.* 84:412–414.

Bergersen, F. J. 1984. Oxygen and the physiology of diazotrophic microorganisms. *In* Veeger, C., and Newton, W. E. (eds.), *Advances in Nitrogen Fixation Research*. Martinus Nijhoff, The Hague, pp. 171–180 (Advances in Agricultural Biotechnology).

Bergersen, F. J., and Turner, G. L. 1980. Properties of terminal oxidase systems of bacteroids from root nodules of soybean and cowpea and of N_2-fixing bacteria grown in continuous culture. *J. Gen. Microbiol.* 118:235–252.

Berlier, Y. M., and Lespinat, P. A. 1980. Mass-spectrometric kinetic studies of the nitrogenase and hydrogenase activities in *in vivo* cultures of *Azospirillum brasilense* Sp7. *Arch. Microbiol.* 125:67–72.

Bhuvaneswari, T. V., Turgeon, G., and Bauer, W. D. 1980. Early events in the infection of soybean (*Glycine max* L. Merr.) by *Rhizobium japonicum*. I. Localization of infectible root cells. *Plant Physiol.* 66:1027–1031.

Boddey, R. M., and Döbereiner, J. 1982. Association of *Azospirillum* and other diazotrophs with tropical gramineae. *In: Non-symbiotic Nitrogen Fixation and Organic Matter in the Tropics*. Indian Society of Soil Science, New Delhi, pp. 28–47.

Boddey, R. M., and Döbereiner, J. 1984. Nitrogen fixation associated with grasses and cereals. *In* Subba Rao, N. S. (ed.), *Current Developments in Biological Nitrogen Fixation*. Oxford and IBH, New Delhi, pp. 277–313.

Boddey, R. M., and Victoria, R. L. 1986. Estimation of biological nitrogen fixation associated with *Brachiaria* and *Paspalum* grasses using ^{15}N labelled organic matter and fertilizer. *Plant Soil* 90:265–292.

Boddey,R. M., Quilt, P., and Ahmad, N. 1978. Acetylene reduction in the rhizosphere of rice: Methods of assay. *Plant Soil* 50:567–574.

Boddey, R. M., Chalk, P. M., Victoria, R. L., and Matsui, E. 1983. The ^{15}N-isotope dilution technique applied to the estimation of biological nitrogen fixation associated with *Paspalum notatum* cv. batatais in the field. *Soil Biol. Biochem.* 15:25–32.

Boddey, R. M., Chalk, P. M., Victoria, R. L., and Matsui, E. 1984. Nitrogen fixation by nodulated soybean under tropical field conditions estimated by the ^{15}N isotope dilution technique. *Soil Biol. Biochem.* 16:593–589.

Boddey, R. M., Baldani, V. L. D., Baldani, J. I., and Döbereiner, J. 1986. Effect of inoculation of *Azospirillum* spp. on the nitrogen assimilation of field grown wheat. *Plant Soil* 95:109–121.

Bopaiah, B. M., and Rai, P. V. 1979. Effect of four common herbicides on the population of microorganisms in red sandy soil. *Plant Soil* 52:451–452.

Bothe, H., Klein, B., Stephan, M. P., and Döbereiner, J. 1981. Transformations of inorganic nitrogen by *Azospirillum* spp. *Arch. Microbiol.* 130:96–100.

Bothe, H., Kronenberg, A., Stephan, M. P., Zimmer, W., and Neuer, G. 1983. Nitrogen fixation and denitrification by a wheat–*Azospirillum* association. *In* Klingmüller, W. (ed.), Azospirillum *II: Genetics, Physiology, Ecology*. Birkhäuser, Basel, pp. 100–113 (Experientia supplementum, 48).

Boulter, D., Jeremy, J. J., and Wilding, M. 1966. Amino acids liberated into the culture medium by pea seedling roots. *Plant Soil* 24:121–127.

Bouton, J. H., and Zuberer, D. A. 1979. Response of *Panicum maximum* Jacq. to inoculation with *Azospirillum brasilense. Plant Soil* 52:585–590.

Bouton, J. H., and Brooks, C. O. 1982. Screening pearl millet for variability in supporting bacterial acetylene reduction activity. *Crop Sci.* 22:680–681.

Bouton, J. H., Smith, R. L., Schank, S. C., Burton, G. W., Tyler, M. E., Littell, R. C., Gallaher, R. N., and Quesenberry, K. H. 1979. Response of pearl millet inbreds and hybrids to inoculation with *Azospirillum brasilense. Crop Sci.* 19:12–16.

Bouton, J. H., Albrecht, S. L., and Zuberer, D. A. 1985. Screening and selection of pearl millet for root associated bacterial nitrogen fixation. *Field Crop Res.* 11:131–140.

Boyle, C. D., and Patriquin, D. G. 1980. Endorhizal and exorhizal acetylene-reducing activity in a grass (*Spartina alterniflora* Loisel.)–diazotroph association. *Plant Physiol.* 66:276–280.

Boyle, C. D., and Patriquin D. G. 1981. Carbon metabolism of *Spartina alterniflora* Loisel in relation to that of associated nitrogen-fixing bacteria. *New Phytol.* 89:275–288.

Bradshaw-Rouse, J. J., Whatley, M. H., Coplin, D. L., Woods, A., Sequeira, L., and Kelman, A. 1981. Agglutination of *Erwinia stewartii* strains with a corn agglutinin: correlation with extracellular polysaccharide production and pathogenicity. *Appl. Environ. Microbiol.* 42:344–350.

Brenchley, J. E. 1973. Effect of methionine sulfoximine and methionine sulfone on glutamate synthesis in *Klebsiella aerogenes. J. Bacteriol.* 114:666–673.

Brewin, N. J., DeJong, T. M., Phillips, D. A., and Johnston, A.W. B. 1980. Cotransfer of determinants for hydrogenase activity and nodulation ability in *Rhizobium leguminosarum. Nature* 288:77–79.

Brown, M. E. 1961. Stimulation of streptomycin resistant bacteria in the rhizosphere of leguminous plants. *J. Gen. Microbiol.* 24:369–377.

Brown, M. E. 1974. Seed and root bacterization. *Annu. Rev. Phytopathol.* 12:181–197.

Brown, M. E., and Burlingham, S. K. 1968. Production of plant growth substances by *Azotobacter chroococcum. J. Gen. Microbiol.* 53:135–144.

Brown, M. E., Burlingham, S. K., and Jackson, R. M. 1962. Studies on *Azotobacter* species in soil. II. Populations of *Azotobacter* in the rhizosphere and effects of artificial inoculation. *Plant Soil* 17:320–332.

Buchanan, R. E. and Gibbons, N. E. 1974. *Bergey's Manual of Determinative Bacteriology.* 8th ed. Williams and Wilkins, Baltimore, 1268 p.

Buchanan-Wollaston, V., Cannon, M. C., Beynon, J. L., and Cannon, F. C. 1981. Role of the nifA gene product in the regulation of nif expression in *Klebsiella pneumoniae. Nature* 294:776–778.

Burgess, B. K. 1984. Structure and reactivity of nitrogenase—an overview. *In: Advances in Nitrogen Fixation Research.* Veeger, C. and Newton, W. E. (eds.). Nijhoff Junk Publishers, The Hague, pp. 103–114.

Burgess, B. K. 1985. Nitrogenase mechanism—an overview. *In: Nitrogen Fixation Research Progress.* Evans, M. J., Bottomly, P. J., Newton, W. E. (eds.). Martinus Nijhoff Publishers, The Hague, pp. 543–549.

Burris, R. H., Albrecht, S. L., and Okon, Y. 1978. Physiology and biochemistry of *Spirillum lipoferum. In* Döbereiner, J, Burris, R. H., Hollaender, A., Franco, A. A., Neyra, C. A., and Scott, D. B. (eds.), *Limitations and Potentials for Biological Nitrogen Fixation in the Tropics.* Plenum, New York, pp. 306–316 (Basic Life Sciences, 10).

Butz, R. G., Long, R. C. 1979. L-malate as an essential component of the xylem fluid of corn seedling roots. *Plant Physiol.* 64:684–689.

Cambraia, J., Galvani, F. R. Estevão, M. M., and Sant'Anna, R. 1983. Effects of aluminum on organic acid, sugar and amino acid composition of the root system of sorghum (*Sorghum bicolor* L.) Moench. *J. Plant Nutr.* 6:313–322.

Campêlo, A. B., and Döbereiner, J. 1970. Occorrência de *Derxia* sp. em solos de alguns estados brasileiros. *Pesq. Agropec. Bras.* 5:327–332.

Cannon, F., Beynon, J., Buchanan-Wollaston, V., Burghoff, R., Cannon, M., Kwiatkowski, R., Lauer, G., and Rubin, R. 1985. Progress in understanding organization and expression of *nif* genes in *Klebsiella. In* Evans, H. J., Bottomley,

P. J., and Newton, W. E. (eds.), *Nitrogen Fixation Research Progress*. Martinus Nijhoff, Dordrecht, pp. 453–460.

Chan, Y. K., Nelson, L. M., and Knowles, R. 1980. Hydrogen metabolism of *Azospirillum brasilense* in nitrogen-free medium. *Can. J. Microbiol.* 26:1126–1131.

Charyulu, P. B. B. N., Ramakrishna, C., and Rao, V. R. 1980. Effect of 2-aminobenzimidazole on nitrogen fixers from flooded soil and their nitrogenase activity. *Bull. Environ. Contam. Toxicol.* 25:482–486.

Charyulu, P. B. B. N., Fourcassie, F., Barbouche, A. K., Rondro-Harisoa, L., Omar, A. M. N., Weinhard, P., Marie, R., and Balandreau, J. 1985. Field inoculation of rice using in vitro selected bacterial and plant genotypes. *In* Klingmüller, W. (ed.), Azospirillum *III: Genetics, Physiology, Ecology*. Springer-Verlag, Berlin, pp. 163–179.

Child, J. J., and Kurz, W. G. W. 1978. Inducing effect of plant cells on nitrogenase activity by *Spirillum* and *Rhizobium* in vitro. *Can. J. Microbiol.* 24:143–148.

Christiansen-Weniger, C., Boddey, R. M., and Döbereiner, J. 1985. Evaluation of nitrogen fixation in sorghum cultivars inoculated with different strains of *Azospirillum* spp. *In* Klingmüller, W. (ed.), Azospirillum *III: Genetics, Physiology, Ecology*. Springer-Verlag, Berlin, pp. 180–188.

Clara, R. W., and Knowles, R. 1984. Superoxide dismutase, catalase, and peroxidase in ammonium grown and nitrogen-fixing *Azospirillum brasilense. Can. J. Microbiol.* 30:1222–1228.

Cohen, E., Okon, Y., Kigel, J., Nur, I., and Henis, Y. 1980. Increase in dry weight and total nitrogen content in *Zea mays* and *Setaria italica* associated with nitrogen-fixing *Azospirillum* spp. *Plant Physiol.* 66:746–749.

Crawford, R. M. M. 1976. Mineral nutrition. *In* Hall, M. A. (ed.), *Plant Structure, Function and Adaptation*. MacMillan, London, pp. 197–253.

Daniel, R. M., Smith, I. M., Phillip, J. A. D., Ratcliffe, H. D., Drozd, J. W., and Bull, A. T. 1980. Anaerobic growth and denitrification by *Rhizobium japonicum* and other rhizobia. *J. Gen. Microbiol.* 120:517–521.

Danneberg, G., Zimmer, W., and Bothe, H. 1985. Some physiological and biochemical properties of denitrification by *Azospirillum brasilense. In* Klingmüller, W. (ed.), Azospirillum *III: Genetics, Physiology, Ecology*. Springer-Verlag, Berlin, pp. 127–138.

Dart, P. J. 1974. The infection process. *In* Quispel, A. (ed.), *The Biology of Nitrogen Fixation*. North-Holland Publishing Co., Amsterdam, pp. 381–429.

Dart, P. J. 1986. Nitrogen fixation associated with nonlegumes in agriculture. *Plant Soil* 90:303–334.

Dart, P. J., and Wai, S. P. 1982. Non-symbiotic nitrogen fixation and soil fertility. *In: Non-symbiotic Nitrogen Fixation and Organic Matter in the Tropics; Symposia Papers 1*. Indian Society of Soil Science, New Delhi, pp. 3–27.

Dart, P. J., Day, J. M., and Harris, D. 1972. Assay of nitrogenase activity by acetylene reduction. *In: Use of Isotopes for Study of Fertilizer Utilization by Legume Crops*. International Atomic Energy Agency, Vienna, pp. 85–100.

Das, A., and Mishra, A. K. 1982. Effect of yeast extract, casamino acids peptone and various L-amino acids, on growth and acetylene reduction in *Azospirillum brasilense. Indian J. Exp.Biol.* 20:751–755.

Das, A., and Mishra, A. K. 1983. Utilization of fructose by *Azospirillum brasilense*. *Can. J. Microbiol.* 29:1213–1217.

Date, R. A. 1971. Microbiological problems in the inoculation of legumes. *Plant Soil* 32:703–725.

Day, J. M., and Döbereiner, J. 1976. Physiological aspects of N_2 fixation by a *Spirillum* from Digitaria roots. *Soil Biol. Biochem.* 8:45–50.

Day, J. M., Harris, D., Dart, P. J., and Van Berkum, P. 1975. The Broadbalk experiment. An investigation of nitrogen gains from non-symbiotic nitrogen fixation. *In* Stewart, W. D. P. (ed.), *Nitrogen Fixation by Free-living Microorganisms*. Cambridge, London, pp. 71–84 (International Biological Programme, 6).

Dazzo, F. B. 1980. Determination of host specificity in the *Rhizobium* clover symbiosis. *In* Newton, W. E., and Orme-Johnson, W. H. (eds.), *Nitrogen Fixation*, Vol. II. University Park, Baltimore, pp. 165–168.

Dazzo, F. B., and Hubbell, D. H. 1975. Antigenic differences between infective and noninfective strains of *Rhizobium trifolii*. *Appl. Environ. Microbiol.* 30:172–177.

Dazzo, F. B., and Milam, J. R. 1975. Serological studies of *Sprillum lipoferum*. *Soil Crop Sci. Soc. Fl. Proc.* 35:121–126.

De Bruijn, F. J., Sundaresan, V.,Szeto, W. W., Ow, D. W., and Ausubel, F. M. 1984. Regulation of the nitrogen fixation (*nif*) genes of *Klebsiella pneumoniae* and *Rhizobium meliloti*: Role of nitrogen regulation (*ntr*) genes. *In* Veeger, C., and Newton, W. E. (eds.), *Advances in Nitrogen Fixation Research*. Martinus Nijhoff, The Hague, pp. 627–633 (Advances in Agricultural Biotechnology).

Dekhuijzen, H. M. 1976. Endogenous cytokinins in healthy and diseased plants. *In* Heitefuss, R., and Williams, P. H. (eds.), *Physiological Plant Pathology*. Springer-Verlag, Berlin, pp. 526–559.

Del Gallo, M. M., Gratani, L., and Morpugo, G. 1985. Mutation in *Azospirillum brasilense*. *In* Klingmüller, W. (ed.), Azospirillum *III: Genetics, Physiology, Ecology*. Springer-Verlag, Berlin, pp. 85–97.

De-Polli, H., and Roth, I. L. 1981. Scanning electron microscopy of large cells of *Azospirillum* species. Annual Meeting of the American Society for Microbiology, Texas, p. 85.

De-Polli, H., Matsui, E., Döbereiner, J., and Salati, E. 1977. Confirmation of nitrogen fixation in two tropical grasses by $^{15}N_2$ incorporation. *Soil Biol. Biochem* 9:119–123.

De-Polli, H., Bohlool, B. B., and Döbereiner, J. 1980. Serological differentiation of *Azospirillum* species belonging to different host-plant specificity groups. *Arch. Microbiol.* 126:217–222.

De-Polli, H., Boyer, C. H., and Neyra, C. A. 1982. Nitrogenase activity associated with roots and stems of field-grown corn (*Zea mays* L.) plants. *Plant Physiol.* 70:1609–1613.

Derx, H. G. 1950. *Beijerinckia*, a new genus of nitrogen-fixing bacteria occurring in tropical soils. *Proc. Kon. Ned. Akad. v. Wet. Amsterdam* 53:140–147.

Derx, H. G. 1953. Sur les causes de la distribution geographique limitee des Beijerinckia. Proc. VI Int. Cong. Microbiol. Roma 6:354–355.

Desale, A. G., and Konde, B. K. 1984. Response of sorghum to seed bacterization with nitrogen levels. *J. Maharastra agr. Univ.* 9:169–170.

De Smedt, J., Bauwens, M., Tytgat, R., and De Ley, J. 1980. Intra- and intergeneric similarities of ribosomal ribonucleic acid cistrons of free-living, nitrogen-fixing bacteria. *Int. J. Syst. Bacteriol.* 30:106–122.

Dilworth, M. J. 1966. Acetylene reduction by nitrogen fixing preparations from *Clostridium pasteurianum. Biochem. Biophys. Acta* 127:285–294.

Diniz, S. P. S. S. 1981. Regulacão da sintese da catalase em *Azospirillum brasilense.* Universidade Federal do Paraná, Curitiba, Brazil. M. Sc. Thesis.

Dixon, R. A. 1984. The genetic complexity of nitrogen fixation. *J. Gen. Microbiol.* 130:2745–2755.

Dixon, R. A., and Postgate, J. R. 1972. Genetic transfer of nitrogen fixation from *Klebsiella pneumoniae* to *Escherichia coli. Nature* 237:102–103.

Dixon, R. A., Cannon,F. C., and Kondorosi, A. 1976. Construction of a P plasmid carrying nitrogen fixation genes from *Klebsiella pneumoniae. Nature* 260:268–271.

Dixon, R. O. D. 1972. Hydrogenase in legume root nodule bacteroids: occurrence and properties. *Arch. Mikrobiol.* 85:193–201.

Döbereiner, J. 1959. Influência da cana-de-acúcar na populacao de *Beijerinckia* no solo. *Rev. bras. Biol* 19:251–258.

Döbereiner, J. 1961. Nitrogen-fixing bacteria of the genus *Beijerinckia Derx* in the rhizosphere of sugar cane. *Plant Soil* 15:211–216.

Döbereiner, J. 1966. *Azotobacter paspali* sp.n., uma bactéria fixadora de nitrogênio na rizosfera de *Paspalum. Pesq. Agropec. Bras.* 1:357–365.

Döbereiner, J. 1968. Non-symbiotic nitrogen fixation in tropical soils. *Pesq. Agropec. Bras.* 3:1–6.

Döbereiner, J. 1970. Further research on *Azotobacter paspali* and its variety specific occurrence in the rhizosphere of *Paspalum notatum* Flügge. *Zentbl. Bakt. Parasitke.* 124:224–230.

Döbereiner, J. 1978a. Influence of environmental factors on the occurrence of *Spirillum lipoferum* in soils and roots. *In* Granhall, U. (ed.), *Environmental Role of Nitrogen-Fixing Blue-Green Algae and Asymbiotic Bacteria.* Swedish Natural Science Research Council, Stockholm, pp. 343–352 (Ecological Bulletins, 26).

Döbereiner, J. 1978b. Nitrogen fixation in grass-bacteria associations in the tropics. *In: Isotopes in Biological Dinitrogen Fixation.* International Atomic Energy Agency, Vienna, pp. 51–69.

Döbereiner, J. 1980. Forage grasses and grain crops. *In* Bergersen, F. J. (ed.), *Methods for Evaluating Biological Nitrogen Fixation.* John Wiley and Sons, Chichester, pp. 535–556.

Döbereiner, J. 1983. Ten years of *Azospirillum. In* Klingmüller, W. (ed.), Azospirillum *II: Genetics, Physiology, Ecology.* Birkhäuser, Basel, pp. 9–23 (Experientia supplementum, 48).

Döbereiner, J., and Baldani, J. I. 1982. Bases científicas para uma agricultura biológica. *Cien. Cult.* 34:869–881.

Döbereiner, J., and Baldani, V. L. D. 1979. Selective infection of maize roots by streptomycin-resistant *Azospirillum lipoferum* and other bacteria. *Can. J. Microbiol.* 25:1264–1269.

Döbereiner, J., and Baldani, V. L. D. 1981. Prospects for inoculation of grasses with *Azospirillum* spp. *In* Vose, P. B., Ruschel, A. P. (eds.), *Associative N_2-Fixation, V. II.* Boca Raton, pp. 33–37.

Döbereiner, J., and Campelo, A. B. 1971. Non-symbiotic nitrogen fixing bacteria in tropical soils. *Plant Soil*, Special Volume:457–470.

Döbereiner, J., and Castro,A. F. 1955. Occorrência e capacidade de fixacão de nitrogênio de bactérias do gênero *Beijerinckia* nas séries de solos da area territorial do Centro Nacional de Ensino e Pesquisas Agronômicas. *Bol. IEEA* 16:1–18.

Döbereiner, J., and Day, J. M. 1975. Nitrogen fixation in the rhizosphere of tropical grasses. *In* Stewart, W. D. P. (ed.), *Nitrogen Fixation by Free-Living Microorganisms.* Cambridge, London, pp. 39–56 (International Biological Programme, 6).

Döbereiner, J., and Day, J. M. 1976. Associative symbioses in tropical grasses: Characterization of microorganisms and dinitrogen fixing sites. *In* Newton, W. E., and Nyman, C. J. (eds.), *Proceedings of the 1st International Symposium on Nitrogen Fixation, V. II.* Washington State University Press, Pullman, pp. 518–538.

Döbereiner, J., and De-Polli, H. 1980. Diazotrophic rhizocoenoses. *In* Stewart, W. D. P. (ed.), *Nitrogen Fixation.* Academic, London, pp. 301–333.

Döbereiner, J., and De-Polli, H. 1981. Nitrogen fixing rhizocoenoses. *In* Russell, R. S., Igue, K., and Mehta, Y. R. (eds.), *The Soil/Root System in Relation to Brazilian Agriculture.* Fundacão Instituto Agronômico do Paraná, Londrina, pp. 175–198.

Döbereiner, J., and Ruschel, A. P. 1958. Uma nova espécie de *Beijerinckia. Rev. Biol.* 1:261–272.

Döbereiner, J., Day, J. M., and Dart, P. J. 1972. Nitrogenase activity and oxygen sensitivity of the *Paspalum notatum-Azotobacter paspali* association. *J. Gen. Microbiol* 71:103–116.

Döbereiner, J., Marriel, I. E., and Nery, M. 1976. Ecological distribution of *Spirillum lipoferum* Beijerinck. *Can. J. Microbiol.* 22:1464–1473.

Dommergues, Y. R., and Rinaudo, G. 1979. Factors affecting N_2 fixation in the rice rhizosphere. *In: Nitrogen and Rice.* Int. Rice Res. Inst., Laguna, pp. 241–260.

Douglas, W. J., Jr., Suslow, T. V., and Steinback, K. E. 1985. Relationship between rapid form adhesion and long term colonization of roots by bacteria. *Appl. Environ. Microbiol.* 50:392–397.

Duncan, M. J. 1979. L-arabinose metabolism in rhizobia. *J. Gen. Microbiol.* 113:177–179.

Eady, R. R. 1981. Regulation of nitrogenase activity. *In* Gibson, A. H., and Newton, W. E. (eds.), *Current Perspectives in Nitrogen Fixation.* Australian Academy of Science, Canberra, pp. 172–181.

Eady, R. R., and Postgate, J. R. 1974. Nitrogenase. *Nature* 249:805–810.

Eady, R. R., Issack, R., Kennedy, C., Postgate, J. R., and Ratcliffe, H. D. 1978. Nitrogenase synthesis in *Klebsiella pneumoniae*: Comparison of ammonium and oxygen regulation. *J. Gen. Microbiol.* 104:227–285.

Ela, S. W., Anderson, M. A., and Brill, W. J. 1982. Screening and selection of maize to enhance associative bacterial nitrogen fixation. *Plant Physiol.* 70:1564–1567.

Elmerich, C. 1983. *Azospirillum* genetics. *In* Puhler, A. (ed.), *Molecular Genetics of the Bacteria-Plant Interaction*. Springer-Verlag, Berlin, pp. 367–372.

Elmerich, C., and Franche, C. 1982. Azospirillum genetics: Plasmids, bacteriophages and chromosome mobilization. *In* Klingmüller, W. (ed.), Azospirillum: *Genetics, Physiology, Ecology*. Birkhäuser, Basel, pp. 9–17 (Experienta supplementum, 42).

Elmerich, C., Quiviger, B., Rosenberg, C., Franche, C., Laurent, P., and Döbereiner, J. (1982) Characterization of a temperate bacteriophage for *Azospirillum*. *Virology* 122:29–37.

Elmerich, C., Fogher, C., Bozouklian, H., Perroud, B., and Dusha, I. 1985. Advances in the genetics of *Azospirillum*. *In* Evans, H. J., Bottomley, P. J., and Newton, W. E. (eds.), *Nitrogen Fixation Research Progress*. Martinus Nijhoff, Dordrecht, pp. 477–483.

Emerich, D. W., and Burris, R. H. 1978. Complementary functioning of the component proteins of nitrogenase from several bacteria. *J. Bacteriol.* 134:936–943.

Eskew, D. L., and Ting, I. P. 1977. Comparison of intact plant and excised root assays for acetylene reduction in grass rhizospheres. *Plant Sci. Lett.* 8:327–331.

Eskew, D. L., Focht, D. D., and Ting, I. P. 1977. Nitrogen fixation, denitrification and pleomorphic growth in a highly pigmented *Spirillum lipoferum*. *Appl. Environ. Microbiol.* 34:592–585.

Eskew, D. L., Eaglesham, A. R. J., and App, A. A. 1981. Heterotrophic $^{15}N_2$ fixation and distribution of newly fixed nitrogen in a rice-flooded soil system. *Plant Physiol.* 68:48–52.

Espin, G., Alvarez-Morales, A., and Merrick, M. 1981. Complementation analysis of *gln A* linked mutations which affect nitrogen fixation in *Klebsiella pneumoniae*. *Mol. Gen. Genet.* 184:213–217.

Evans, H. J., Purohit, K., Cantrell, M., Eisbrenner, G., Russel, S. A., Hanus, F. J., and Lepo, J. E. 1981. Hydrogen losses and hydrogenases in nitrogen-fixing organisms. *In* Gibson, A., and Newton, W. E. (eds.), *Current Perspectives in Nitrogen Fixation*. Australian Academy of Science, Canberra, pp. 84–96.

Fahsold, R., Singh, M., and Klingmüller, W. 1985. Cosmid cloning of nitrogenase structural genes of *Azospirillum lipoferum*. *In* Klingmüller, W. (ed.), Azospirillum *III: Genetics, Physiology, Ecology*. Springer–Verlag, Berlin, pp. 30–40.

Falk, E. C., Döbereiner, J., Johnson, J. L., and Krieg, N. R. 1985. Deoxyribonucleic acid homology of *Azospirillum amazonense* Magalhães *et al.* 1984 and emendation of the description of the genus *Azospirillum*. *Int. J. Syst. Bacteriol.* 35:117–118.

Falk, E. C., Johnson, J. L., Baldani, V. L. D., Döbereiner, J., and Krieg, N. R. 1986. Deoxyribonucleic and ribonucleic acid homology studies of the genera *Azospirillum* and Conglomeromonas. *Int. J. Syst. Bacteriol.* 36:80–85.

Fani, R., Bazzicalupo, M., Gallori, E., Turbanti, L., and Polsinelli, M. 1985. Construction of a gene bank of *Azospirillum brasilense*. *In* Klingmüller, W. (ed.), Azospirillum *III: Genetics, Physiology, Ecology*. Springer–Verlag, Berlin, pp. 1–9.

Fayez, M., Eman, N. F., and Makbould, H. E. 1983. Interaction of the herbicides bromoxynil and afalon S with *Azospirillum* and growth of maize. *Z. Pflanzenern Bodenkd.* 146:741–751.

Ferreira, M. C. B., Fernandes, M. S., Döbereiner, J. 1987. Role of *Azospirillum* nitrate reductase in nitrate assimilation by wheat plants. *Biol. Fert. Soils* (in press).

Fogher, C., Bozouklian, H., Bandhari, S. K., and Elmerich, C. 1985a. Construction of a genomic library of *Azospirillum brasilense* Sp7 and cloning of the glutamine synthetase gene. *In* Klingmüller, W. (ed.), Azospirillum *III: Genetics, Physiology, Ecology*. Springer–Verlag, Berlin, pp. 41–51.

Fogher, C., Dusha, I., Barbot, P., and Elmerich, C. 1985b. Heterologous hybridization of *Azospirillum* DNA to *Rhizobium nod* and *fix* genes. *FEMS Microbiol. Lett.* 30:245–250.

Fogher, C., Bozouklian, H., Bhandari, S., and Elmerich, C. 1985c. Cloning and characterization of the *gln A* gene of *Azospirillum brasilense. In* Evans, M. J., Bottomley, P. J., and Newton, W. E. (eds.), *Nitrogen Fixation Research Progress.* Martinus Nijhoff, Dordrecht, p. 521.

Foster, T. J. 1983. Plasmid-determined resistance to antimicrobial drugs and toxic metal ions in bacteria. *Microbiol. Rev.* 47:361–409.

Franche, C., and Elmerich, C. 1981. Physiological properties and plasmid content of several strains of *Azospirillum brasilense* and *A. lipoferum. Ann. Microbiol.* 132:3–18.

Franche, C., Canelo, E., Gauthier, D., and Elmerich, C. 1981. Mobilization of the chromosome of *Azospirillum brasilense* by plasmid R68.45. *FEMS Microbiol. Lett.* 10:199–202.

Freitas, J. L. M. de, Rocha, R. E. M. da, Ferreira, P. A. A., and Döbereiner, J. 1982. Matéria orgânica e inoculacão com *Azospirillum* na incorporacão de N pelo milho. *Pesq. Agropec. Bras.* 17:1423–1432.

Fridovich, I. 1977. Oxygen is toxic. *Bioscience* 27:462–466.

Funayama, S., Rego, L. V., and Pedrosa, F. O. 1985. *Nif*A⁻ mutants of *Azospirillum brasilense* strain Sp245. *In* Evans, H. J., Bottomley, P. J., and Newton, W. E. (eds.), *Nitrogen Fixation Research Progress.* Martinus Nijhoff, Dordrecht, p. 522.

Gallon, J. R. 1981. The oxygen sensitivity of nitrogenase a problem for biochemists and microorganisms. *Trends Biochem. Sci.* 6:19–23.

Gallori, E., and Bazzicalupo, M. 1985. Effect of nitrogen compounds on nitrogenase activity in *Azospirillum brasilense. FEMS Microbiol. Lett.* 28:35–38.

Gardner, J. M., Chandler, J. L., and Feldman, A. W. 1985. Growth responses and vascular plugging of citrus inoculated with rhizobacteria and xylem-resident bacteria. *Plant Soil* 86:333–345.

Gauthier, D., and Elmerich, C. 1977. Relationship between glutamine synthetase and nitrogenase in *Spirillum lipoferum. FEMS Microbiol. Lett.* 2:101–104.

Germida, J. J. 1984. Spontaneous induction of bacteriophage during growth of *Azospirillum brasilense* in complex media. *Can. J. Microbiol.* 30:805–808.

Gibson, J. E., Stackebrandt, E., Zablen, L. B., Gupta, R., and Woese, C. R. 1979. A phylogenetic analysis of the purple photosynthetic bacteria. *Curr. Microbiol.* 3:59–64.

Giddens, J. 1977. Nitrogen accumulation by grasses. *Ga. Agric. Res.* 19:12–13.

Giles, K. L., and Whitehead, H. C. M. 1975. The transfer of nitrogen-fixing ability to a eukaryote cell. *Cytobios* 14:49–62.

Giller, K. E., Day, J. M., Dart, P. J., and Wani, S. P. 1984. A method for measuring the transfer of fixed nitrogen from free-living bacteria to higher plants using $^{15}N_2$. *J. Microbiol. Meth.* 2:307–316.

Goebel, E. M., and Krieg, N. R. 1984. Fructose catabolism in *Azospirillum brasilense* and *Azospirillum lipoferum*. *J. Bacteriol.* 159:86–92.

Gowda, T. K. S., and Watanabe, I. 1983. Autoradiographic technique for screening chemolithotrophic hydrogen-utilizing bacteria. *Can. J. Microbiol.* 29:1081–1103.

Gupta, K. D., and Ghosh, S. 1984. Identification of a phosphoenolpyruvate: Fructose 1-phosphotransferase system in *Azospirillum brasilense*. *J. Bacteriol.* 160:1204–1206.

Haahtela, K., and Kari, K. 1986. The role of root-associated *Klebsiella pneumoniae* in the nitrogen nutrition of *Poa pratensis* and *Triticum aestivum* as estimated by the method of ^{15}N isotope dilution. *Plant Soil* 90:245–254.

Haahtela, K., and Korhonen, T. K. 1985. *In vitro* adhesion of N_2-fixing enteric bacteria to roots of grasses and cereals. *Appl. Environ. Microbiol* 49(5):1186–1190.

Haahtela, K., Wartiovaara, T., Sundman, V., and Skujins, J. 1981. Root associated N_2 fixation (acetylene reduction) by *Enterobacteriaceae* and *Azospirillum* strains in cold-climate spodosols. *Appl. Environ. Microbiol.* 41:203–206.

Haahtela, K., Helander, I., Nurmiaha-Lassila, E. L., and Sundman, V. 1983a. Morphological and physiological characteristics and lipopolysaccharide composition of N_2-fixing (C_2H_2-reducing) root-associated *Pseudomonas* sp. *Can. J. Microbiol.* 29:874–880.

Haahtela, K., Kari, K., and Sundman, V. 1983b. Nitrogenase activity (acetylene reduction) of root-associated, cold-climate *Azospirillum, Enterobacter, Klebsiella*, and *Pseudmonas* species during growth on various carbon sources and at various partial pressures of oxygen. *Appl. Environ. Microbiol.* 45:563–570.

Haas, D., and Holloway, B. W. 1978. Chromosome mobilization by the plasmid R68.45: A tool in *Pseudomonas* genetics. *Mol. Gen. Gene.* 158:229–237.

Hall, P. G., Krieg, N. R. 1983. Swarming of *Azospirillum brasilense* on solid media. *Can. J. Microbiol.* 29:1592–1594.

Hall, P. G., and Krieg, N. R. 1984. Application of the indirect immunoperoxidase stain technique to the flagella of *Azospirillum brasilense*. *Appl. Environ. Microbiol.* 47:4533–435.

Halliwell, B. 1981. Free radicals, oxygen toxicity and aging. *In* Sohal, R. S. (ed.), *Age Pigments*. Elsevier North-Holland, New York, pp. 1–62.

Halsall, D. M., and Gibson, A. H. 1985. Cellulose decomposition and associated nitrogen fixation by mixed cultures of *Cellulomonas gelida* and *Azospirillum* species or *Bacillus macerans*. *Appl. Environ. Microbiol.* 50:1021–1026.

Halsall, D. M., Turner, G. L., and Gibson, A. M. 1985. Straw and xylem utilization by pure cultures of nitrogen-fixing *Azospirillum* spp. *Appl. Environ. Microbiol.* 49:423–428.

Hardy, K. 1981. *Bacterial Plasmids*. Thomas Nelson and Sons Ltd., Surrey, England.

Hartmann, A. 1982. Antimetabolite effects on nitrogen metabolism of *Azospirillum* and properties of resistant mutants. *In* Klingmüller W. (ed.), Azospirillum: *Genetics, Physiology, Ecology*. Birkhäuser, Basel, pp. 59–68 (Experientia supplementum, 42).

Hartmann, A., and Kleiner, D. 1982. Ammonium (methylammonium) transport by *Azospirillum* spp. *FEMS Microbiol. Lett* 15:65–67.

Hartmann, A., Fußeder, A., and Klingmüller, W. 1983a. Mutants of *Azospirillum* affected in nitrogen fixation and auxin production. *In* Klingmüller, W. (ed.), Azospirillum *II: Genetics, Physiology, Ecology*. Birkhäuser, Basel, pp. 78–88 (Experientia supplementum, 48).

Hartmann, A., Singh, M., and Klingmüller, W. 1983b. Isolation and characterization of *Azospirillum* mutants excreting high amounts of indoleacetic acid. *Can. J. Microbiol.* 29:916–923.

Hartmann, A., Fu, H.-A, Song, S.-D, and Burris, R. H. 1985. Comparison of nitrogenase regulation in *A. brasilense, A. lipoferum* and *A. amazonense. In* Klingmüller, W. (ed.), Azospirillum *III: Genetics, Physiology, Ecology*. Springer–Verlag, Berlin, pp. 116–126.

Hartmann, A., Fu, H.-A, and Burris, R. H. 1986. Regulation of nitrogenase activity by ammonium chloride in *Azospirillum* spp. *J. Bacteriol.* 165:864–870.

Hegazi, N. A., Amer, N. A., Monib, M. 1979. Enumeration of N_2-fixing spirilla. *Soil Biol. Biochem.* 11:437–438.

Hegazi, N. A., Monib, M., Amer, H. A., Shokr, E.-S. 1983. Response of maize plants to inoculation with azospirilla and (or) straw amendment in Egypt. *Can. J. Microbiol* 29:888–894.

Heinrich, D., and Hess, D. 1985. Chemotactic attraction of *Azospirillum lipoferum* by wheat roots and characterization of some attractants. *Can. J. Microbiol.* 31:26–31.

Hess, D., Heinrich,D., Kemmner, J., and Jekel, S. 1985. Secretion and cleavage of sucrose by wheat: A system of chemotactic attraction and nitrogenase induction in *Azospirillum lipoferum. In* Klingmüller, W. (ed.), Azospirillum *III: Genetics, Physiology, Ecology*. Springer–Verlag, Berlin, pp. 147–162.

Heulin, T., Bally, R., and Balandreau, J. 1982. Isolation of a very efficient N_2-fixing bacteria from the rhizosphere of rice. *In* Klingmüller, W. (ed.), Azospirillum: *Genetics, Physiology, Ecology*. Birkhäuser, Basel, pp. 92–99 (Experientia supplementum, 42).

Heulin, T., Weinhard, P., and Balandreau, J. 1983. Motility changes in *Azospirillum lipoferum. In* Klingmüller, W. (ed.), Azospirillum *II: Genetics, Physiology, Ecology*. Birkhäuser, Basel, pp. 89–94 (Experientia supplementum, 48).

Higuti, I. H., and Pedrosa, F. O. 1985. Mutant of *Azospirillum brasilense* Sp7 that overproduces catalase. *In* Evans, H. J., Bottomley, P. J., and Newton, W. E. (eds.), *Nitrogen Fixation Research Progress*. Martinus Nijhoff, Dordrecht, p. 532.

Hill, S. 1971. Influence of oxygen concentration on the colony type of *Derxia gummosa* grown on nitrogen free media. *J. Gen. Microbiol.* 67:77–83.

Hill, S. 1976. Influence of atmospheric oxygen concentrations on acetylene reduction and efficiency of nitrogen fixation in intact *Klebsiella pneumoniae. J. Gen. Microbiol.* 93:335–345.

Hill, S. 1978. Factors influencing the efficiency of nitrogen fixation in free-living bacteria. *In* Granhall, U. (ed.), *Environmental Role of Nitrogen-Fixing Blue-Green Algae and Asymbiotic Bacteria*. Swedish Natural Science Research Council, Stockholm, pp. 130–136 (Ecological Bulletins/NRF 26).

Hill, S. 1985. Redox regulation of enteric *nif* expression is independent of the *fnr* gene product. *FEMS Microbiol. Lett.* 29:5–9.

Hill, W. A., Bacon-Hill, P., Crossman, S. M., and Stevens, C. 1983. Characterization of N_2-fixing bacteria associated with sweet potato roots. *Can. J. Microbiol.* 29:860–862.

Hirota, Y., Fujii, T., Sano, Y., and Iyama, S. 1978. Nitrogen fixation in the rhizosphere of rice. *Nature* 276:416–417.

Holt, J. G., and Krieg, N. R. (eds.). 1984. *Bergey's Manual of Systematic Bacteriology* 9th ed., Vol. I, Williams and Wilkins, Baltimore.

Hubbell, D. H., and Gaskins, M. H. 1984. Associative N_2 fixation with *Azospirillum*. *In* Alexander, M. (ed.), *Biological Nitrogen Fixation*. Plenum Publishing Co., pp. 201–224.

Hurek, T. 1984. Untersuchungen zur Absorbtion Stickstoff fixierender Mikroorganismen an Wurzeln von *Leptochloa fusca* (Linn.) Kunth. Universität Hannover, FRG, 105 p. B. S. Thesis (Diplomarbeit).

Iruthayathas, E. E., Gunasekaran, S., and Vlassak, K. 1983. Effect of combined inoculation of *Azospirillum* and *Rhizobium* on nodulation and N_2-fixation of winged bean and soybean. *Sci. Hortic.* 20:231–240.

Ito, O., Cabrera, D., and Watanabe, I. 1980. Fixation of dinitrogen-15 associated with rice plants. *J. Appl. Environ. Microbiol.* 39:554–558.

Jackson, R. M., and Brown, M. E. 1966. Behaviour of *Azotobacter chroococcum* introduced into the plant rhizosphere. *Ann. Inst. Pasteur* 111 (Suppl.):103–112.

Jagnow, G. 1985. Root colonization and plant growth of grasses and cereals in unsterile soil and nitrogenase (C_2H_2) activity of sterile cereal seedlings after inoculation with *Azospirillum* spp. *In* Klingmüller, W. (ed.), Azospirillum *III: Genetics, Physiology, Ecology*. Springer–Verlag, Berlin, pp. 203–214.

Jain, D. K., and Patriquin, D. G. 1984. Root hair deformation, bacterial attachment, and plant growth in wheat-*Azospirillum* associations. *Appl. Environ. Microbiol.* 48:1208–1213.

Jara, P., Quiviger, B., Laurent, P., and Elmerich, E. 1983. Isolation and genetic analysis of *Azospirillum brasilense* Nif^- mutants. *Can. J. Microbiol.* 29:968–972.

Jensen, H. L. 1965. Nonsymbiotic nitrogen fixation. *In* Bartholomew, W. V., and Clark, F. E. (eds.), *Soil Nitrogen*. American Society of Agronomy, Madison, pp. 436–480 (Agronomy, 10).

Jensen, H. L., Peterson, E. J., De P. K., and Bhattacharya, R. 1960. A new nitrogen-fixing bacterium: *Derxia gummosa* nov. gen. nov. spec. *Arch. Mikrobiol.* 36:182–195.

Johnston, A. W. B., Beynon, J. L., Buchanan-Wollaston, A. V., Setchell, S. M., Hirsch, P. R., and Beringer, J. E. 1978. High frequency transfer of nodulating ability between strains and species of *Rhizobium*. *Nature* 276:634–636.

Kapulnik, Y., Okon, Y., Kigel, J., Nur, I., and Henis, Y. 1981a. Effects of temperature, nitrogen fertilization, and plant age on nitrogen fixation by *Setaria italica* inoculated with *Azospirillum brasilense* (strain Cd). *Plant Physiol.* 68:340–343

Kapulnik, Y., Kigel, J., Okon, Y., Nur, I., and Henis, Y. 1981b. Effect of *Azospirillum* inoculation on some growth parameters and N-content of wheat, sorghum and *Panicum*. *Plant Soil* 61:65–70.

Kapulnik, Y., Sarig, S., Nur, I., Okon, Y., and Henis, Y. 1982. The effect of *Azospirillum* inoculation on growth and yield of corn. *Isr. J. Bot.* 31:247˜55.

Kavimandan, S. K., Subba Rao, N. S., and Mohrir, A. V. 1978. Isolation of *Spirillum lipoferum* from the stems of wheat and nitrogen fixation in enrichment cultures. *Curr. Sci.* 47:96–98.

King, E. O., Ward, M. K., and Raney, D. E. 1954. Two simple media for the demonstration of pyocyanin and fluorescein. *J. Lab. Clin. Med.* 44:301–307.

Kipe-Nolt, J. A., Avalakki, U. K., and Dart, P. J. 1985. Root exudation of sorghum and utilization of exudates by nitrogen-fixing bacteria. *Soil Biol. Biochem.* 17:859–863.

Kleeberger, A., and Klingmüller, W. 1980. Plasmid-mediated transfer of nitrogen-fixing capability to bacteria from the rhizosphere of grasses. *Mol. Gen. Gene.* 180:621–627.

Kloss, M. 1981. Methoden zum Nachweis der Pflanzenwurzelexsudation Organischer Säuren, die Erährungsgrundlage für Stickstoffixierende Bakterien in der Rhizosphäre von *Diplachne fusca* (Linn.) Beauv. bilden. Universität Hannover, 66 p. B. S. Dissertation (Diplomarbeit).

Klucas, R. V., Pedersen, W., Shearman, R. C., and Wood, L. V. 1981. Nitrogen fixation associated with winter wheat, sorghum, and Kentucky bluegrass. *In* Vose P. B., and Ruschel, A. P. (eds.), *Associative N_2-Fixation*, v. I. CRC Press, Boca Raton, pp. 119–129.

Koch, B. L. 1977. Associative nitrogenase activity by some Hawaiian grass roots. *Plant Soil* 47:703–706.

Kolb, W., and Martin, P. 1985. Response of plant roots to inoculation with *Azospirillum brasilense* and to application of indole acetic acid. *In* Klingmüller, W. (ed.), Azospirillum *III: Genetics, Physiology, Ecology* Springer-Verlag, Berlin, pp. 215–221.

Korhonen, T. E., Nurmiaho-Lassila, E.-L., Laakso, T., and Haahtela, K. 1986. Adhesion of fimbriated nitrogen-fixing enteric bacteria to roots of grasses and cereals. *Plant Soil* 90:59–69.

Krasil'nikov, N. A. 1958. *Soil Microorganisms and Higher Plants*. Academy of Sciences, U.S.S.R. English translation available from Office Technical Services, United States Department of Commerce, Washington, D.C.

Krieg, N. R., and Döbereiner, J. 1984. Genus *Azospirillum* Tarrand, Krieg and Döbereiner 1979, 79[A1] (Effective publication: Tarrand, Krieg and Döbereiner 1978, 948). *In* Holt, J. G., and Krieg, N. R. (eds.), *Bergey's Manual of Systematic Bacteriology* 9th ed., v. 1. Williams and Wilkins, Baltimore, pp. 94–104.

Krinsky, N. I. 1979. Carotenoid protection against oxidation. *Pure Appl. Chem.* 51:649–660.

Krishnapillai, V., and Postgate, J. R. 1980. Expression of *Klebsiella his* and *nif* genes in *Serratia marcescens, Erwinia herbicola* and *Proteus mirabilis*. *Arch. Microbiol.* 127:115–118.

Kulasooriya, S. A., Roger, P. A., Barraquio, W. L., and Watanbe, I. 1981. Epiphytic nitrogen fixation in deepwater rice. *Soil Sci. Plant Nutr.* 27:19–27.

Kulinska, D. 1983. Occurrence of *Azospirillum* in Polish soils. *Acta Microbiol. Pol* 32:265–268.

Kulinska, D., and Drozdowicz, A. 1983. Occurrence of microorganisms antagonistic to *Azospirillum* spp. *Zentralbl. Mikrobiol.* 138:585–594.

Ladha, J. K., Barraquio, W. L., and Watanabe, I. 1982. Immunological techniques to identify *Azospirillum* associated with wetland rice. *Can. J. Microbiol.* 28:478–485.

Lakshmi, V., Rao, A. S., Vijayalakshmi, K., Lakshmi-Kumari, M., Tilak, K. V. B. R., and Subba Rao, N. S. 1977. Establishment and survival of *Spirillum lipoferum*. Proc. Indian Acad. Sci. Sect. B 86:397–404.

Lamm, R. B., and Neyra, C. A. 1981. Characterization and cyst production of azospirilla isolated from selected grasses growing in New Jersey and New York. *Can. J. Microbiol.* 27:1320–1325.

Larson, R. I., Neal, J. L. 1978. Selective colonization of the rhizosphere of wheat by nitrogen-fixing bacteria. *In* Granhall, U. (ed.), *Environmental Role of Nitrogen-Fixing Blue-Green Algae and Asymbiotic Bacteria.* Swedish Natural Science Research Council, Stockholm, pp. 331–342 (Ecological Bulletins, 26).

Lee, K. K., Alimagno, B., and Yoshida, T. 1977a. Field technique using the acetylene reduction method to assay nitrogenase activity and its association with the rice rhizosphere. *Plant Soil* 47:519–526.

Lee, K. K., Castro, T., and Yoshida, T. 1977b. Nitrogen fixation throughout growth, and varietal differences in nitrogen fixation by the rhizosphere of rice planted in pots. *Plant Soil* 48:613–619.

Lesinger, T., and Margroff, R. 1979. Secondary metabolites of the fluorescent *Pseudomonas. Microbiol. Rev.* 43:422–442.

Liao, C. F. H. 1981. Devarda's alloy method for total nitrogen determination. *Soil Sci. Soc. Am. J.* 45:852–855.

Lima, E., Boddey, R. M., and Döbereiner, J. 1986. Quantification of biological nitrogen fixation associated with sugar cane using on ^{15}N aided nitrogen balance. *Soil Biol. Biochem.* (submitted).

Lindberg, T., and Granhall, U. 1984. Isolation and characterization of dinitrogen-fixing bacteria from the rhizosphere of temperate cereals and forage grasses. *Appl. Environ. Microbiol.* 48:683–689.

Lindberg, T., Granhall, U., and Tomenius, K. 1985. Infectivity and acetylene reduction of diazotrophic rhizosphere bacteria in wheat (*Triticum aestivum*) seedlings under gnobiotic conditions. *Biol. Fert. Soils* 1:123–129.

Line, M. A., and Loutit, M. W. 1971. Non-symbiotic nitrogen-fixing organisms from some New Zealand tussock-grassland soils. *J. Gen. Microbiol.* 66:309–318.

Loh, W. H. T., Randles, C. I., Sharp, W. R. and Miller, R. H. 1984. Intermediary carbon metabolism of *Azospirillum brasilense. J. Bacteriol.* 158:264–268.

Loureiro, M. F. 1985. Balanço de nitrogênio em gramíneas do gênero *Brachiaria.* Universidade Federal Rural do Rio de Janeiro, Itaquai-RJ, 102 p., M. S. Thesis.

Ludden, P. W., and Burris, R. H. 1976. Activating factor for the iron protein of nitrogenase from *Rhodospirillum rubrum. Science* 194:424–425.

Ludden, P. W., Okon, Y., and Burris, R. H. 1978. The nitrogenase system of *Spirillum lipoferum*. *Biochem. J.* 173:1001–1003.

Macura, J. 1966. Interactions nutritionelles plantes–bacteries et bases experimentales de la bacterization des graines. *Ann. Inst. Pasteur* 111 (suppl.):9–38.

Magalhães, F. M. M., and Döbereiner, J. 1984. Ocorrência de *Azospirillum amazonense* sp. nov., em alguns ecossistemas de Amazônia. *Rev. Microbiol.* 15:246–252.

Magalhães, F. M. M., Patriquin, D., and Döbereiner, J. 1979. Infection of field grown maize with *Azospirillum* spp. *Rev. Bras. Biol.* 39:587–596.

Magalhães, F. M. M., Baldani, J. I., Souto, S. M., Kuykendall, J. R., and Döbereiner, J. 1983. A new acid-tolerant *Azospirillum* species. *An. Acad. Bras. Ciên.* 55:417–430.

Magalhães, L. M. S., Neyra, C. A., and Döbereiner, J. 1978. Nitrate and nitrite reductase negative mutants of N_2-fixing *Azospirillum* spp. *Arch. Microbiol.* 117:247–252.

Malik, K. A., and Schlegel, H. G. 1981. Chemolithoautotrophic growth of bacteria able to grow under N_2-fixing conditions. *FEMS Microbiol. Lett.* 11:63–67.

Malik, K. A., and Zafar, Y. 1985. Quantification of root associated nitrogen fixation in kallar grass as estimated by ^{15}N isotope dilution. *In* Malik, K. A., Naqvi, S. H., and Aleem, M. I. H. (eds.), *Nitrogen and the Environment*. Nuclear Institute for Agriculture and Biology, Faisalabad, pp. 161–171.

Manga, V. K., Venkateswarlu, B., and Saxena, M. B. L. 1985. Gene action for nitrogenase activity in the roots of pearl millet. *Indian J. Agric. Sci.* 55:391–393.

Marocco, A., Bazzicalupo, M., and Perenzin, M. 1983. Forage grasses inoculation with gentamicine and sulfaguanidine resistant mutants of *Azospirillum brasilense*. *In* Klingmüller, W. (ed.), Azospirillum *II: Genetics, Physiology, Ecology*. Birkhäuser, Basel, pp. 149–158 (Experientia supplementum, 48).

Marriel, I. E., and Cruz, J. C. 1978. Increased N_2 fixation (C_2H_2) in field grown maize by herbicide treatments. *In: Limitations and Potentials for Biological Nitrogen Fixation in the Tropics*. Plenum Press, New York, p. 340 (Basic Life Sciences, 10).

Marschner, M. 1985. Nährstoff Dynamik in der Rhizosphäre. *Ber. Deutsch Bot. Ges.* 98:291–309.

Martin, J. K. 1975. ^{14}C-labelled material leached from the rhizosphere of plants supplied continuously with $^{14}CO_2$. *Soil Biol. Biochem.* 7:395–399.

Martin, J. K. 1977. Factors affecting the loss of organic carbon from wheat roots. *Soil Biol. Biochem.* 9:1–8.

Martin, P., Glatzle, A. 1982. Mutual influences of *Azospirillum* spp. and grass seedlings. *In* Klingmüller, W. (ed.), Azospirillum *II: Genetics, Physiology, Ecology*. Birkhäuser, Basel, pp. 108–220 (Experientia supplementum, 48).

Martinez-Drets, G., Del Gallpo, M., Burpee, C., and Burris, R. H. 1984. Catabolism of carbohydrates and organic acids and expression of nitrogenase by *azospirilla*. *J. Bacteriol.* 159:80–85.

Martinez-Drets, G., Fabiano, E., and Cardona, A. 1985. Carbohydrate catabolism in *Azospirillum amazonense*. *Appl. Environ. Microbiol.* 50:183–185.

Martinez-Toledo, M. V., Gonzalez-Lopez, J., Rubia, T. de la, and Ramos-Cormenzana, A. 1985. Isolation and characterization of *Azotobacter chroococcum* from the roots of *Zea mays*. *FEMS Microbiol. Ecol.* 31:197–203.

Matthysse, A. G., Hohne, K. V., and Gurlitz, R. H. G. 1981. Elaboration of cellulose fibrils by *Agrobacterium tumefaciens* during attachment to carrot cells. *J. Bacteriol* 145:583–595.

Matsubara, T. 1970. Studies on denitrification. *J. Biochem* 67:229–235.

McAuliffe, C., Chamblee, D. S., Uribe-Arango, H., and Woodhouse, W. W. 1958. Influence of inorganic nitrogen on nitrogen fixation by legumes as revealed by ^{15}N. *Agron. J.* 50:334–347.

McClung, C. R., and Patriquin, D. G. 1980. Isolation of a nitrogen-fixing *Campylobacter* species from the roots of *Spartina alterniflora* Loisel. *Can. J. Microbiol.* 26:881–886.

Merrick, M. J. 1983. Nitrogen control of the *nif* regulon in *Klebsiella pneumoniae*: Involvement of the *ntr*A gene and analogies between *ntr*C and *nif*A. *EMBO J.* 2:39–44.

Merrick, M.J., and Dixon, R. 1984. Why don't plants fix nitrogen? *Trends Biotechnol.* 2:162–166.

Michiels, K., Vanstockem, M., Vanderleyden, J., and Van Gool, A. 1985. Stability of broad host range plasmids in *Azospirillum* cloning of a 59 KBP plasmid from *A. brasilense* RO7. *In* Klingmüller, W. (ed.), Azospirillum *III: Genetics, Physiology, Ecology*. Springer-Verlag, Berlin, pp. 63–73.

Millet, E., Avivi, Y., and Feldman, M. 1985. Effects of rhizospheric bacteria on wheat yield under field conditions. *Plant Soil* 86:347–355.

Minchin, F. R., Witty, J. F., Sheehy, J. E., and Müller, M. 1983. A major error in the acetylene reduction assay: Decreases in nodular nitrogenase activity under assay conditions. *J. Exp. Bot.* 34:641–649.

Miranda, C. H. B., and Boddey, R. M. 1986. Estimation of biological nitrogen fixation associated with 11 ecotypes of *Panicum maximum* grown in ^{15}N labelled soil. *Agron. J.* (in press).

Miranda, C. H. B., Seiffert, N. F., and Döbereiner, J. 1985. Efeito de aplicacão de molibdênio no número de *Azospirillum* e na producão de *Brachiaria decumbens*. *Pesq. Agropec. Bras.* 20:509–513.

Mishra, A. K., and Roy, P. 1981. Mobilisation of *Spirillum lipoferum* chromosomal DNA by RP4. *Indian J. Microbiol* 21:60–62.

Mishra, A. K., Roy, P., and Bhattacharya, S. 1979. Deoxyribonucleic acid-mediated transformation of *Spirillum lipoferum*. *J. Bacteriol.* 137:1425–1427.

Moore, A. W. 1966. Non-symbiotic nitrogen fixation in soil and soil plant systems. *Soils Fert.* 29:113–128.

Moore, E. R. B., Morrod, E. P., and Jurtshuk, P., Jr. 1984. Superoxide dismutases of *Azotobacter vinelandii* and other aerobic free-living nitrogen-fixing bacteria. *FEMS Microbiol. Lett.* 24:261–265.

Morris, D. R., Zuberer, D. A., and Weaver, R. W. 1985. Nitrogen fixation by intact grass-soil cores using $^{15}N_2$ and acetylene reduction. *Soil Biol. Biochem.* 17:87–91.

Mortenson, L. E., and Thorneley, R. N. F. 1979. Structure and function of nitrogenase. *Ann. Rev. Biochem.* 48:387–418.

Murty, M. G. 1984. Phyllosphere of cotton as a habitat for diazotrophic microorganisms. *Appl. Environ. Microbiol.* 46:713–718.

Nair, S. K., Jara, P., Quiviger, B., and Elmerich, C. 1983. Recent developments in the genetics of nitrogen fixation in *Azospirillum. In* Klingmüller, W. (ed.), Azospirillum *II: Genetics, Physiology, Ecology*. Birkhäuser, Basel, pp. 29–38 (Experientia supplementum, 48).

Napoli, C., Dazzo, F., and Hubbell, D. 1975. Production of microfibrils by *Rhizobium. Appl. Microbiol.* 30:123–131.

Nayak, D. N., and Rao, V. R. 1977. Nitrogen fixation by *Spirillum* sp. from rice roots. *Arch. Microbiol* 115:359–360.

Nayak, D. N., Swain, A., and Rao, V. R. 1979. Nitrogen-fixing *Azospirillum lipoferum* from common weeds associated with rice and aquatic ecosystems. *Curr. Sci.* 48:866–867.

Neal, J. L., Jr. and Larson, R. I. 1976. Acetylene reduction by bacteria isolated from the rhizosphere of wheat. *Soil Biol. Biochem.* 8:151–155.

Nelson, A. D., Barber, L. E., Tjepkema, J., Russell, S. A., Powelson, R., Evans, H. J., and Seidler, R. J. 1976. Nitrogen fixation associated with grasses in Oregon. *Can. J. Microbiol.* 22:523–530.

Nelson, L. M., and Knowles, R. 1978. Effect of oxygen and nitrate on nitrogen fixation and denitrification by *Azospirillum brasilense* grown in continuous culture. *Can. J. Microbiol* 24:1395–1403.

Nery, M., Abrantes, G. T. V.,Santos, O. dos, and Döbereiner, J. 1977. Fixacão de nitrogênio em trigo. *R. Bras. Ci. Solo.* 1:15–20.

Neuer, G., Kronenberg, A., and Bothe, H. 1985. Denitrification and nitrogen fixation by *Azospirillum*. III. Properties of a wheat-*Azospirillum* association. *Arch. Microbiol* 141:364–370.

Neves, M. C. P., Nery, M., and Day, J. M. 1976. Efeito da temperatura na fixação de nitrogênio de estirpes de *Spirillum* sp. isoladas de *Digitaria* sp. e milho. *In: Congresso Brasileiro De Ciências Do Solo*, 15. Anais, Campinas-SP. pp. 143–149.

Neyra, C. A., and Van Berkum, P. 1976. Nitrogenase activity in isolated sorghum roots: Effects of bicarbonate and inorganic nitrogen. *Plant Physiol.* 57(Suppl.):533.

Neyra, C. A., and Van Berkum, P. 1977. Nitrate reduction and nitrogenase activity in *Spirillum lipoferum. Can. J. Microbiol.* 23:306–310.

Neyra, C. A., Döbereiner, J., Lalande, R., and Knowles, R. 1977. Denitrification by N_2-fixing *Spirillum lipoferum. Can. J. Microbiol.* 23:300–305.

Nioh, I. 1979. Nitrogen fixation and a nitrogen-fixing bacterium from the roots of *Eragrostis ferruginéa. J. Gen. Appl. Microbiol.* 25 :261–271.

Novick, N. J., and Tyler, M. E. 1982. L-arabinose metabolism in *Azospirillum brasilense. J. Bacteriol.* 149:364–367.

Novick, N. J., and Tyler, M. E. 1983. Partial purification and properties of an L-arabinose dehydrogenase from *Azospirillum brasilense. Can. J. Microbiol.* 29:242–246.

Nur, I., Okon, Y., and Henis, Y. 1980a. An increase in nitrogen content of *Setaria italica* and *Zea mays* inoculated with *Azospirillum. Can. J. Microbiol.* 26:482–485.

Nur, I., Okon, Y., and Henis, Y. 1980b. Comparative studies of nitrogen-fixing bacteria associated with grasses in Israel with *Azospirillum brasilense*. *Can. J. Microbiol.* 26:714–718.

Nur, I., Steinitz, Y. L., Okon, Y., and Henis, Y. 1981. Carotenoid composition and function in nitrogen-fixing bacteria of the genus *Azospirillum*. *J. Gen. Microbiol.* 122:27–32.

Nur, I., Okon, Y., and Henis, Y. 1982a. Effect of dissolved oxygen tension on production of carotenoids, poly-β-hydroxybutyrate, succinate oxidase and superoxide dismutase by *Azospirillum brasilense*-Cd grown in continuous culture. *J. Gen. Microbiol.* 128:2937–2943.

Nur, I., Okon, Y., and Henis, Y. 1982b. The role of oxygen concentration in the synthesis of carotenoids, poly-β-hydroxybutyrate and succinate oxidase in continuous culture of *Azospirillum*. *Isr. J. Bot.* 31:332–227.

Nurmiaho-Lassila, E. L., Haahtela, K., and Sundman, V. 1981. A new spiral structure associated with the flagellum of *Azospirillum lipoferum*. *Can. J. Microbiol.* 27:1267–1271.

Nuti, M. P., Lepidi, A. A., Prakash, R. K., Schilperoort, R. A., and Cannon, F. C. 1979. Evidence for nitrogen fixation (*nif*) genes on indigenous *Rhizobium* plasmids. *Nature* 282:533–535.

Odom, J. M., and Peck, H. D., Jr. 1984. Hydrogenase, electron-transfer proteins and energy coupling in the sulfate-reducing bacteria *Desulfovibrio*. *Ann. Rev. Microbiol.* 38:551–592.

O'Hara, G. W., Davey, M. R., and Lucas, J. A. 1981. Effect of inoculation of *Zea mays* with *Azospirillum brasilense* strains under temperate conditions. *Can. J. Microbiol.* 27:871–877.

O'Hara, G. W., Davey, M. R., and Lucas, J. A. 1983. Association between the nitrogen fixing bacterium *Azospirillum brasilense* and excised plant roots. *Z. Pflanzenphysiol.* 113:1–13.

Okon, Y. 1982. *Azospirillum*: Physiology, properties, mode of association with roots and its application for the benefit of cereal and forage grass crops. *Isr. J. Bot.* 31:214–220.

Okon, Y. 1985. *Azospirillum* as a potential inoculant for agriculture. *Trends Biotechnol.* 3:223–228.

Okon, Y., Albrecht, S. L., and Burris, R. H. 1976a. Factors affecting growth and nitrogen fixation of *Spirillum lipoferum*. *J. Bacteriol.* 127:1248–1254.

Okon, Y., Albrecht, S. L., and Burris, R. H. 1976b. Carbon and ammonia metabolism of *Spirillum lipoferum*. *J. Bacteriol.* 128:592–597.

Okon, Y., Albrecht, S. L., and Burris, R. H. 1977a. Methods for growing *Spirillum lipoferum* and for counting it in pure culture and in association with plants. *Appl. Environ. Microbiol.* 33:85–88.

Okon, Y., Houchins, J. P., Albrecht, S. L., and Burris, R. H. 1977b. Growth of *Spirillum lipoferum* at constant partial pressures of oxygen, and the properties of its nitrogenase in cell-free extracts. *J. Gen. Microbiol.* 98:87–93.

Okon, Y., Cakmakci, L., Nur, I., and Chet, I. 1980. Aerotaxis and chemotaxis of *Azospirillum brasilense*. *Microb. Ecol.* 6:277–280.

Okon, Y., Nur, I., and Henis, Y. 1983. Effect of oxygen concentration on electron transport components and microaerobic properties of *Azospirillum brasi-*

lense. *In* Klingmüller, W. (ed.), Azospirillum *II: Genetics, Physiology, Ecology*. Birkhäuser, Basel, pp. 115–126 (Experientia supplementum, 48).

Oliveira, R. G. B., and Drozdowicz, A. 1981. Bacteriocins in the genus *Azospirillum*. *Rev. Microbiol.* 12:42–47.

O'Toole, P., and Knowles, R. 1973. Efficiency of acetylene reduction (nitrogen fixation) in soil: Effect of type and concentration of available carbohydrate. *Soil Biol. Biochem* 5:789–797.

Ow, O. W., and Ausubel, F. M. 1983. Regulation of nitrogen metabolism genes by the *nif*A gene product in *Klebsiella pneumoniae*. *Nature* 301:307–313.

Page, W. J. 1985. Genetic transformation of molybdenum-starved *Azotobacter vinelandii*: Increased transformation frequency and recipient range. *Can. J. Microbiol* 31:659–662.

Panopoulos, N. J., Walton, J. D., and Willis, D. K. 1984. Genetic and biochemical basis of virulence in plant pathogens. *In* Verma, D. P. S., and Hohn, Th. (eds.), *Genes Involved in Microbe-Plant Interaction*. Springer-Verlag, Vienna, pp. 339–374.

Papen, H., and Werner, D. 1982. Organic acid utilization, succinate excretion and encystation during biphasic nitrogenase activity in *Azospirillum brasilense* during microaerobic conditions. *In* Klingmüller, W. (ed.), Azospirillum *II: Genetics, Physiology, Ecology*. Birkhäuser, Basel, pp. 75–91 (Experientia supplementum, 42).

Parker, C. A. 1957a. Non-symbiotic nitrogen-fixing bacteria in soil. III. Total nitrogen changes in a field soil. *J. Soil Sci.* 8:48–59.

Parker, C. A. 1957b. Evolution of nitrogen fixing symbioses in higher plants. *Nature* 179:593–594.

Patriquin, D. G. 1978. Factors affecting nitrogenase activity (acetylene reducing activity) associated with excised roots of the emergent halophyte *Spartina alterniflora* Loisel. *Aquat. Bot.* 4:193–210.

Patriquin, D. G. 1981. New developments in grass–bacteria associations. *In* Subba Rao, N. S. (ed.), *Advances in Agricultural Microbiology*. Oxford and IBH Publishing, New Delhi, pp. 139–190.

Patriquin, D. G., and Denike, D. 1978. *In situ* acetylene reduction assays of nitrogenase activity associated with the emergent halophyte *Spartina alterniflora* Loisel: Methodological Problems. *Aquat. Bot.* 4:211–226.

Patriquin, D. G., and Döbereiner, J. 1978. Light microscopy observations of tetrazolium-reducing bacteria in the endorhizosphere of maize and other grasses in Brazil. *Can. J. Microbiol.* 24:734–742.

Patriquin, D. G., and McClung, C. R. 1978. Nitrogen accretion, and the nature and possible significance of N_2 fixation (acetylene reduction) in a Nova Scotia *Spartina alterniflora* stand. *Mar. Biol.* 47:227)n242.

Patriquin, D. G., Gracioli, L. A., and Ruschel, A. P. 1980. Nitrogenase activity of sugar cane propagated from stem cuttings in sterile vermiculite. *Soil Biol. Biochem.* 12:413–417.

Patriquin, D. G., Döbereiner, J., and Jain, D. K. 1983. Sites and processes of association between diazotrophs and grasses. *Can. J. Microbiol.* 29:900–915.

Pedersen, W. L., Chakrabarty, K., Klucas, R. V., and Vidaver, A. K. 1978. Nitrogen fixation (acetylene reduction) associated with roots of winter wheat and sorghum in Nebraska. *Appl. Environ. Microbiol.* 35:129–135.

Pedrosa, F. O., and Yates, M. G. 1983a. Effect of chelating agents and nickel ions on hydrogenase activity in *Azospirillum brasilense, A. lipoferum* and *Derxia gumosa. FEMS Microbiol. Lett.* 17:101–106.

Pedrosa, F. O., and Yates, M. G. 1983b. *Nif* mutants of *Azospirillum brasilense*: Evidence for a *nif*A type regulation. *In* Klingmüller, W. (ed.), Azospirillum *II: Genetics, Physiology, Ecology*. Birkhäuser, Basel, pp. 66–77 (Experientia supplementum, 48).

Pedrosa, F. O., and Yates, M. G. 1984. Regulation of nitrogen fixation (*nif*) genes of *Azospirillum brasilense* by *nif*A and *ntr(gln)* type gene products. *FEMS Microbiol. Lett.* 23:95–101.

Pedrosa, F. O., Döbereiner, J., and Yates, M. G. 1980. H_2-dependent growth and autotrophic CO_2 fixation by *Derxia. J. Gen. Microbiol.* 119:547–551.

Pedrosa, F. O., Stephan, M. P., and Döbereiner, J. 1981. Interaction of nitrogenase and uptake hydrogenase activities in *Azospirillum brasilense. In* Vose, P. B., and Ruschel, A. P. (eds.), *Associative Nitrogen Fixation*, v. I. CRC, Boca Raton Press, pp. 15–25.

Pedrosa, F. O., Stephan, M. P., Döbereiner, J., and Yates, M. G. 1982. Hydrogen-uptake hydrogenase activity in nitrogen-fixing *Azospirillum brasilense. J. Gen. Microbiol.* 128:161–166.

Peña, J. J., and Döbereiner, J. 1974. Efecto del nitrato y amonio en la actividad de la nitrogenase de bactérias tropicales fijadoras de nitrógeno atmosférico. *R. Lat. Microbiol.* 16:33–42.

Pereira, P. A. A., Von Bülow, J. F. W., and Neyra, C. A. 1978. Atividade da nitrogenase, nitrato redutase e acumulacão de nitrogênio em milho braquítico *Zea mays* L. (cv. Piranão) em dois níveis de adubacão nitrogenda. *Rev. Bras. Cien. Solo* 2:28–33.

Pereira, P. A. A., Döbereiner, J., and Neyra, C. A. 1981. Nitrogen assimilation and dissimilation in five genotypes of *Brachiaria* spp. *Can. J. Microbiol.* 59:1475–1479.

Perroud, B., Bandhari, S. K., and Elmerich, C. 1985. The *nif*HDK operon of *Azospirillum brasilense* Sp7. *In* Klingmüller, W. (ed.), Azospirillum *III: Genetics, Physiology, Ecology*. Springer-Verlag, Berlin, pp. 10–19.

Plazinski, J., and Rolfe, B. G. 1985a. Interaction of *Azospirillum* and *Rhizobium* strains leading to inhibition of nodulation. *Appl. Environ. Microbiol* 49: 990–993.

Plazinski, J., and Rolfe, B. G. 1985b. Influence of *Azospirillum* strains on the nodulation of clovers by *Rhizobium* strains. *Appl. Environ. Microbiol.* 49:984–989.

Plazinski, J., and Rolfe, B. G. 1985c. *Azospirillum-Rhizobium* interaction leading to a plant-growth stimulation without nodule formation. *Can. J. Microbiol.* 31:1026–1030.

Plazinski, J., Dart, P. J., and Rolfe, B. G. 1983. Plasmid visualization and *nif* gene location in nitrogen-fixing *Azospirillum* strains. *J. Bacteriol* 155:1429–1433.

Pohlman, A. A., and McColl, J. G. 1983. Nitrogen fixation in the rhizosphere and rhizoplane of barley. *Plant Soil* 69:341–352.

Polsinelli, M., Baldanzi, E., Bazzicaluppo, M., and Gallori, E. 1980. Transfer of plasmid pRD1 from *Escherichia coli* to *Azospirillum brasilense. Mol. Gen. Genet.* 178:709–711.

Postgate, J. R. 1971. *The Chemistry and Biochemistry of Nitrogen Fixation.* Plenum, London, 316 p.

Postgate, J. R. 1982. *The Fundamentals of Nitrogen Fixation.* Cambridge University, Cambridge, 252 p.

Postgate, J. R., and Kent, H. 1985. Expression of *Klebsiella pneumoniae nif* genes in *Proteus mirabilis. Arch. Microbiol.* 142:289–294.

Postgate, J. R., and Krishnapillai, V. 1977. Expression of *Klebsiella nif* and *his* genes in *Salmonella typhimurium. J. Gen. Microbiol.* 98:379–385.

Purchase, B. S. 1978. Nitrogen fixation associated with grasses: A potential source of nitrogen for Rhodesian agriculture. *Rod. Agric. J.* 75:99–104.

Quiviger, B., Franche, C., Lutfalla, G., Haselkorn, R., and Elmerich, C. 1982. Cloning of a nitrogen fixation (*nif*) gene cluster of *Azospirillum brasilense. Biochimie* 64:495–502.

Raju, P. N., Evans, H. J., and Seidler, R. J. 1972. An asymbiotic nitrogen-fixing bacterium from the root environment of corn. *Proc. Natl. Acad. Sci. USA* 69:3474–3478.

Rao, A. V., and Venkateswarlu,B. 1985. Most probable numbers of *Azospirillum* associated with the roots of inoculated pearl millet. *Plant Soil* 88:153–158.

Ratti, S., Curti, B., Zanetti, G., and Galli, E. 1985. Purification and characterization of glutamate synthase from *Azospirillum brasilense. J. Bacteriol.* 163:724–729.

Reiner, O., Okon, Y., and Eisenbach, M. 1985. Oxygen recognition in aerotactic behaviour of *Azospirillum brasilense. In* Evans, H. J., Bottomley, P. J., and Newton, W. E. (eds.), *Nitrogen Fixation Research Progress.* Martinus Nijhoff, Dordrecht, p. 426.

Reinhold, B. Hurek, T., and Fendrik, I. 1985. Strain specific chemotaxis of *Azospirillum* spp. *J. Bacteriol.* 162:190–195.

Reinhold, B., Hurek, T., Fendrik, I., Pot, B., Gillis, M., Kertsers, K., Thielemans, D., and De Ley, J. 1987. *Azospirillum halopraeferans* sp. nov., a nitrogen fixing organism associated with roots of Kallar grass [*Leptochloa fusca* (L. Kunth]. *Int. J. Syst. Bacteriol.* 37:43–51.

Reinhold, B., Hurek, T., Niemann, E. G., and Fendrik, I. 1986. Close association of *Azospirillum* and diazotrophic rods with different root zones of Kallar grass. *Appl. Environm. Microbiol.* 52:520–526.

Rennie, R. J. 1980a. Dinitrogen-fixing bacteria: computer assisted identification of soil isolates. *Can. J. Microbiol.* 26:1275–1283.

Rennie, R. J. 1980b. ^{15}N-isotope dilution as a measure of dinitrogen fixation by *Azospirillum brasilense* associated with maize. *Can. J. Bot.* 58:21–24.

Rennie, R. J., and Larson, R. I. 1979. Dinitrogen fixation associated with disomic chromosome substitution lines of spring wheat. *Can. J. Bot.* 57:2771–2775.

Rennie, R. J., and Larson, R. I. 1981. Dinitrogen fixation associated with disomic chromosome substitution lines of spring wheat in the phytotron and in the field. *In* Vose, P. B., and Ruschel, A. P. (eds.), *Associative N_2-fixation*, v. II. CRC Press, Boca Raton, pp. 145–154.

Rennie, R. J., De Freitas, J. R., Ruschel, A. P., and Vose, P. B. 1982. Isolation and identification of N_2-fixing bacteria associated with sugar cane (*Saccharum* sp.). *Can. J. Microbiol.* 28:462–467.

Reynders, L., and Vlassak, K. 1979. Conversion of tryptophan to indole-acetic acid by *Azospirillum brasilense. Soil Biol. Biochem.* 11:547–548.

Reynders, L., and Vlassak, K. 1982. Use of *Azospirillum brasilense* as biofertilizer in intensive wheat cropping. *Plant Soil* 66:217–223.

Riedel, G. E., Ausubel, F. M., and Cannon, F. C. 1979. A physical map of chromosomal nitrogen fixation (*nif*) genes of *Klebsiella pneumoniae. Proc. Natl. Acad. Sci. USA* 76:2866–2870.

Rinaudo, G. 1977. La fixation d'azote dans la rhizosphère du riz: Importance du type varieétal. *Cah. ORSTOM, Ser. Biol.* 12:117–119.

Robson, R. L., and Postgate, J. R. 1980. Oxygen and hydrogen in biological nitrogen fixation. *Ann. Rev. Microbiol* 34:183–207.

Robson, R. L., Kennedy, C., and Postgate, J. R. 1983. Progress in comparative genetics of nitrogen fixation. *Can. J. Microbiol* 29:954–967.

Rocha, R. E. M., Baldani, J. I., and Döbereiner, J. 1981. Specificity of infection by *Azospirillum* spp. in plants with C_4 photosynthetic pathway. *In* Vose, P. B., and Ruschel, A. P. (eds.), *Associative N_2-fixation*, v. II. CRC Press, Boca Raton, pp. 67–69.

Rovira, A. D. 1963. Microbial inoculation of plants. I. Establishment of free living nitrogen fixing bacteria in the rhizosphere and their effects on maize, tomato and wheat. *Plant and Soil* 19:304–314.

Rovira, A. D. 1965. Interaction between plant roots and soil microorganisms. *Ann. Rev. Microbiol* 19:241–266.

Rovira, A. D. 1969. Diffusion of carbon compounds away from wheat roots. *Aust. J.Biol. Sci.* 22:1287–1290.

Ruschel, A. P., and Döbereiner, J. 1965. Bactérias assimbióticas fixadoras de nitrogênio na rizosfera de gramíneas forrageiras. *In: An. IX Congresso Internacional de Pastagens*, São Paulo, v. 2, pp. 1103–1107.

Ruscoe, A. W., Newcomb, E. H., and Burris, R. H. 1978. Pleomorphic forms in strains of *Spirillum lipoferum* by light and electron microscopy. *In* Proceedings of the Steenbock-Kettering International Symposium on Nitrogen Fixation, Madison, Wisconsin, Abst. 60.

Sadasivan, L., and Neyra, C. A. 1985. Flocculation of *Azospirillum brasilense* and *Azospirillum lipoferum*: Exopolysaccharides and cyst formation. *J. Bacteriol.* 163:716–723.

Sampaio, M. J. A. M., Silva, E. M. R., Döbereiner, J., Yates, M. G., and Pedrosa, F. O. 1981. Autotrophy and methylotrophy in *Derxia gummosa, Azospirillum brasilense* and *A. lipoferum. In* Gibson, A. H., and Newton, W. E. (eds.), *Current Perspectives in Nitrogen Fixation.* Australian Academy of Science, Canberra, p. 444.

Sampaio, M. J. A. M., Pedrosa, F. E., and Döbereiner, J. 1982. Growth of *Derxia gummosa* and *Azospirillum* spp. on C1-compounds. *An. Acad. Bras. Cien.* 54:457–458.

Schank, S. C., Smith, R. L., Weiser, G. C., Zuberer, D. A., Bouton, J. H., Quesenberry, K. H., Tyler, M. E., Milam, J. R., and Littell, R. C. 1979. Fluorescent antibody technique to identify *Azospirillum brasilense* associated with roots of grasses. *Soil Biol. Biochem.* 11:287–295.

Schmidt, E. L. 1979. Initiation of plant root-microbe interactions. *Ann. Rev. Microbiol.* 33:355–376.

Schöllhorn, R., and Burris, R. H. 1966. Study of intermediates in nitrogen fixation. *Fed. Proc.* 25:710.

Schröder, M. 1932. Die Assimilation des Luftstickstoffs durch einige Bakterien. *Zentralbl. Bakt. Parasitenkd.* 85:178–212.

Schwabe, G., and Weber, U. 1983. Uptake and fate of plasmid-DNA in *Azospirillum lipoferum* A23. *In* Klingmüller, W. (ed.), Azospirillum *II: Genetics, Physiology, Ecology*. Birkhäuser, Basel, pp. 56–65 (Experientia supplementum, 48).

Schwabe, G., Posseckert, G., and Klingmüller, W. 1985. Restriction endonucleases in *Azospirillum. Gene* 39:113–116.

Scott, C. A., Magalhães, F. M. M., Divan, V. L. dos S., and Scott, D. B. 1978. Numbers of *Azospirillum* spp. associated with the roots of field grown maize. *In* Döbereiner, J., Burris, R. H., Hollaender, A., Franco, A. A., Neyra, C. A., and Scott, D. B. (eds.), *Limitations and Potentials for Biological Nitrogen Fixation in the Tropics*. Plenum Press, New York, pp. 371–372 (Basic Life Sciences, 10).

Scott, D. B., Scott, C. A., and Döbereiner, J. 1979. Nitrogenase activity and nitrate respiration in *Azospirillum* spp. *Arch. Microbiol.* 121:141–145.

Scowcroft, W. R., and Gibson, A. H. 1975. Nitrogen fixation by *Rhizobium* associated with tobacco and cowpea cell cultures. *Nature* 253:351–352.

Seldin, L., and Dubnau,D. 1985. Deoxyribonucleic acid homology among *Bacillus polymyxa, Bacillus macerans, Bacillus azotofixans*, and other nitrogen-fixing *Bacillus* strains. *Int. J. Syst. Bacteriol.* 35:151–154.

Seldin, L., Van Elsas, J. D., Penido, E. G. C. 1984. *Bacillus azotofixans* sp. nov. a nitrogen-fixing specie from Brazilian soils and grass roots. *Int. J. Syst. Bacteriol.* 34:451–456.

Sequeira, L., and Graham, T. L. 1977. Agglutination of avirulent strains of *Pseudomonas solanacearum* by potato lectin. *Physiol. Plant Pathol.* 11:43–54.

Shimschick, E. J., and Hebert, R. R. 1979. Binding characteristics of N_2-fixing bacteria to cereal roots. *Appl. Environ. Microbiol.* 38:447–453.

Silveira-Schrank, I., and Drozdowicz, A. 1983. The behaviour of *Azospirillum* spp. in the presence of plant growth hormones. *Ann. Microbiol. (Inst. Pasteur)* 134A:411–414.

Simpson, F. R., and Burris, R. H. 1984. A nitrogen pressure of 50 atmospheres does not prevent evolution of hydrogen by nitrogenase. *Science* 224:1095–1097.

Singh, C. S., and Subba Rao, N. S. 1979. Associative effect of *Azospirillum brasilense* with *Rhizobium japonicum* on nodulation and yield of soybean (*Glycine max*). *Plant Soil* 53:387–392.

Singh, M. 1982. Transfer of bacteriophage Mu and transposon Tn5 into *Azospirillum. In* Klingmüller, W. (ed.), Azospirillum: *Genetics, Physiology, Ecology*. Birkhäuser, Basel, pp. 35–43 (Experientia supplementum, 42).

Singh, M., and Klingmüller, W. 1985. Problems and prospects of site-directed transposon mutagenesis. *In* Klingmüller, W. (ed.), Azospirillum *III: Genetics, Physiology, Ecology*. Springer-Verlag, Berlin, pp. 20–29.

Singh, M., and Wenzel, W. 1982. Detection and characterication of plasmids in *Azospirillum. In* Klingmüller, W. (ed.), Azospirillum: *Genetics, Physiology, Ecology*. Birkhäuser, Basel, pp. 44–51 (Experientia supplementum, 42).

Singh, M., Kleeberger, A., and Klingmüller, W. 1983. Location of nitrogen fixation (*nif*) genes on indigenous plasmids of *Enterobacter agglomerans*. *Mol. Gen. Genet.* 190:373–378.

Singh, S., Ganguly, T. K., Neelakatan, S., Singh, K., Singh, A., and Tomer, P. S. 1980. Note on the response of sorghum and cowpea to *Azospirillum brasilense* and nitrogen. *Indian J. Agric. Sci.* 50:721–724.

Skerman, V. B. D., Sly, L. I., and Williamson, M. L. 1983. *Conglomeromonas largomobilis* gen. nov., sp. nov., a sodium-sensitive, mixed-flagellated organism from fresh waters. *Int. J. Syst. Bacteriol.* 33:300–308.

Smith, R. L., Bouton, J. H., Schank, S. C., Quesenberry, K. H., Tyler, M. e., Milam, J. R., Gaskins, M. H., and Littell, R. C. 1976. Nitrogen fixation in grasses inoculated with *Spirillum lipoferum*. *Science* 193:1003–1005.

Smith, R. L., Schank, s. C., Milam, J. R., and Baltensperger, A. A. 1984. Responses of *Sorghum* and *Pennisetum* species to the N_2-fixing bacterium *Azospirillum brasilense*. *Appl. Environ. Microbiol.* 47:1331–1336.

Socolofsky, M. D., and Wyss, O. 1962. Resistance of the *Azotobacter* cyst. *J. Bacteriol.* 84:119–124.

Song, S. D., Hartmann, A., and Burris, R. H. 1985. Purification and properties of the nitrogenase of *Azospirillum amazonense*. *J. Bacteriol.* 164:1271–1277.

Souto, S. M., and Döbereiner, J. 1971. Fixação do nitrogênio atmosferico por *Beijerinckia* na rizosfera do capim elefante-napier de Pinda (*Pennisetum purpureum*). *In* An. XI Congresso Brasileiro de Ciencias do Solo, Brasîlia, pp. 32–33.

Souto, S. M., and Döbereiner, J. 1984. Metodologia para medição da fixação de N_2 em raízes extraídas de gramíneas forrageiras tropicais. *Pesq. Agropec. Bras.* 19:553–565.

Souto, S. M., and Döbereiner, J. 1985. Variação estacional da fixação de N_2 e assimilação de nitrato em gramíneas forageiras tropicais. *Pesq. Agropec. Bras.* 20:319–334.

Sprent, J. I. 1979. *The Biology of Nitrogen-Fixing Organisms*. McGraw-Hill, London, 196 p.

Starkey, R., and De, P. K. 1939. A new species of *Azotobacter*. *Soil Sci.* 47:329–343.

Stephan, M. P., Pedrosa, F. O., and Döbereiner, J. 1981. Physiological studies with *Azospirillum* spp. *In* Vose, P. B., and Ruschel, A. P. (eds.), *Associative N_2-Fixation*, v. I. CRC Press, Boca Raton, pp. 7–13.

Stephan, M. P., Zimmer, W., and Bothe, H. 1984. Denitrification by *Azospirillum brasilense* Sp7. II. Growth with nitrous oxide as respiratory electron acceptor. *Arch. Microbiol.* 138:212–216.

Strzelczykova, A., and Strzelczyk, E. 1961. Effect of different carbon and nitrogen sources on the antagonistic activity against *Azotobacter* and *Rhizobium* in various strains of soil streptomycetes. *Acta Microbiol. Pol.* 10:101–107.

Stumpf, D. K., Russell, M., and Albrecht, S. L. 1977. N_2 fixation by *Spirillum lipoferum* in association with crop plants, and the relationship of organic soils. *Plant Physiol.* 59:95.

Subba Rao, N. S. 1981. Response of crops to *Azospirillum* inoculation in India. *In* Vose, P. B., and Ruschel,A. P. (eds.), *Associative N_2-Fixation*, v. I. CRC Press, Boca Raton, pp. 137–144.

Subba Rao, N. S. 1983. Nitrogen-fixing bacteria associated with plantation and orchard plants. *Can. J. Microbiol.* 29:863–873.

Subba Rao, N. W., Tilak, K. V. B. R., and Singh, C.S. 1985a. Synergistic effect of vesicular-arbuscular mycorrhizas and *Azospirillum brasilense* on the growth of barley in pots. *Soil Biol. Biochem.* 17:119–121.

Subba Rao, N. S., Tilak, K. V. B. R., and Singh, C. S. 1985b. Effect of combined inoculation of *Azospirillum brasilense* and vesicular-arbuscular mycorrhiza on pearl millet. *Plant Soil* 84:283–286.

Tal, S., and Okon, Y. 1985. Production of the reserve material poly-β-hydroxybutyrate and its function in *Azospirillum brasilense* Cd. *Can. J. Microbiol.* 31:608–613.

Tarrand, J. J., Krieg, N. R., and Döbereiner, J. 1978. A taxonomic study of the *Spirillum lipoferum* group with description of a new genus, *Azospirillum* gen. nov. and two species, *Azospirillum lipoferum* (Beijerinck) comb. nov. and *Azospirillum brasilense* sp. nov. *Can. J. Microbiol.* 24:967–980.

Tavares, M. C. D. 1978. Efeito dos herbicidas no equilíbrio biológico do solo de cerrado. Instituto de Microbiologia da UFRJ, Rio de Janeiro, 114 p. Ph.D. Thesis.

Taylor, R. W. 1979. Response of 2 grasses to inoculation with *Azospirillum* spp. in a Bahamian soil. *Trop. Agr.* 56:361–365.

Tchan, Y. T. 1957. Studies of nitrogen-fixing bacteria. Vi. A new species of nitrogen-fixing bacteria. *Proc. Linn. Soc., N. S. W.* 82:314–316.

Thomas-Bauzon, D., Weinhard, P., Villecourt, P., and Balandreau, J. 1982. The spermosphere model. I. Its use in growing, counting, and isolating N_2-fixing bacteria from the rhizosphere of rice. *Can. J. Microbiol.* 28:922–928.

Thorneley, R. N. F., and Eady, R. R. 1977. Nitrogenase of *Klebsiella pneumoniae* distinction between proton-reducing and acetylene reducing forms of the enzyme: effect of temperature and component ration on substrate-reduction kinetics. *Biochem. J.* 167:457–461.

Tibelius, K. H., and Knowles, R. 1983. Effect of hydrogen and oxygen on uptake-hydrogenase activity in nitrogen-fixing and ammonium-grown *Azospirillum brasilense. Can. J. Microbiol.* 29:1119–1125.

Tibelius, K. H., and Knowles, R. 1984a. Uptake hydrogenase activity in denitrifying *Azospirillum brasilense* grown anaerobically with nitrous oxide or nitrate. *J. Bacteriol.* 157:84–88.

Tibelius, K. H., and Knowles, R. 1984b. Hydrogenase activity in *Azospirillum brasilense* is inhibited by nitrite, nitric oxide, carbon monoxide, and acetylene. *J. Bacteriol.* 160:103–106.

Tien, T. M., Gaskins, M. H., and Hubbell, D. H. 1979. Plant growth substances produced by *Azospirillum brasilense* and their effect on the growth of pearl millet (Pennisetum americanum L.). *Appl. Environ. Microbiol.* 37:1016–1024.

Tien, T. M., Diem, H. G., Gaskins, M. H., and Hubbell, D. H. 1981. Polygalacturonic acid transeliminase production by *Azospirillum* species. *Can. J. Microbiol.* 27:426–431.

Tjepkema, J. D., and Burris, R. H. 1976. Nitrogenase activity associated with some Wisconsin prairie grasses. *Plant Soil* 45:81–94.

Tjepkema, J. D., and Van Berkum, P. 1977. Acetylene reduction by soil cores of maize and sorghum in Brazil. *Appl. Environ. Microbiol.* 33:626–628.

Trolldenier, G. 1977. Influence of some environmental factors on nitrogen fixation in the rhizosphere of rice. *Plant Soil* 47:203–217.

Tyler, M. E., Milam, J. R., Smith, R. L., Schank, S. C., and Zuberer, D. A. 1979. Isolation of *Azospirillum* from diverse geographic regions. *Can. J. Microbiol.* 25:693–697.

Umali-Garcia, M., Hubbell, D. H., and Gaskins, M. H. 1978. Infection of *Panicum maximum* by *Spirillum lipoferum. In* Granhall, U. (ed.), *Environmental Role of Nitrogen-Fixing Blue-Green Algae and Asymbiotic Bacteria.* Swedish Natural Science Research Council, Uppsala, pp. 373–379 (Ecological Bulletin, 26).

Umali-Garcia, M., Hubbell, D. H., Gaskins, H., and Dazzo, F. B. 1980. Association of *Azospirillum* with grass roots. *Appl. Environ. Microbiol.* 39:219–226.

Uozumi, T., Barraquio, W. L., Wang, P. L., Murai, F., Chung, K. S., and Beppu, T. 1982. Plasmids and *Nif* genes in rhizobia and nitrogen-fixing bacteria in the rhizosphere of rice. *In* Proceedings of the Fourth International Symposium on Genetics of Industrial Organisms. Kyoto, pp. 314–319.

Van Berkum, P., and Bohlool, B. B. 1980. Evaluation of nitrogen fixation by bacteria in association with roots of tropical grasses. *Microb. Rev.* 44:491–517.

Van Berkum, P., and Day, J. M. 1980. Nitrogenase activity associated with soil cores of grasses in Brazil. *Soil Biol. Biochem.* 12:137–140.

Van Berkum, P., and Sloger, C. 1982. Ontogenetic variation of nitrogenase, nitrate reductase, and glutamine synthetase activities in *Oryza sativa. Plant Physiol.* 68:722–726.

Van Berkum, P., and Sloger, C. 1981. Physiology of root-associated nitrogenase activity in *Oryza sativa. Plant Physiol.* 69:1161–1164.

Vancura, V. 1980. Fluorescent pseudomonads in the rhizosphere of plants and their relation to root exudates. *Folia Microbiol.* 25:168–173.

Vancura, V., Abd-El-Malek, Y., and Zayed, M. N. 1965. *Azotobacter* and *Beijerinckia* in the soils and rhizosphere of plants in Egypt. *Folia Microbiol.* 10:224–229.

Vancura, V., Macura, I., Fischer, O., and Vandracek, J. 1959. The relation of *Azotobacter* in the root systems of barley. *Folia Microbiol.* 4:119–129.

Van Gundy, S. D., and Freckman, D. W. 1977. Phytoparasitic nematodes in below ground agroecosystems. *Eco. Bull.* (Stockholm) 23:320–329.

Vanstockem, M., Michiels, K., Vanderleyden, J., Van Good, A. 1985. Transfer and random integration of Tn5 in *Azospirillum. In* Klingmüller, W. (ed.), Azospirillum *III: Genetics, Physiology, Ecology*. Springer-Verlag, Berlin, pp. 74–84.

Vasil, V., Vasil, I. K., Zuberer, D. A., and Hubbell, D. H. 1979. The biology of *Azospirillum*—sugarcane association. I. Establishment of the association. *Z. Pflanzenphysiol.* 95:141–147.

Venkateswarlu, B., and Rao A. V. 1983. Response of pearl millet to inoculation with different strains of *Azospirillum brasilense. Plant Soil* 74:379–386.

Venkateswarlu, B., Rao, A. V. 1985. Interactions between the root exudates of pearl millet and *Azospirillum brasilense. Proc. Indian Nat. Sci. Acad., Part A* 95:237–245.

Vlassak, K., Paul, E. A., and Harris, R. E. 1973. Assessment of biological nitrogen fixation in grassland and associated sites. *Plant Soil* 38:637–649.

Vlassak, K., and Reynders, L. 1978. Association of free-living nitrogen fixing bacteria with plant roots in temperate regions. *In* Loutit, M. W., Miles, J. A. R. (eds.), *Microbial Ecology*. Springer-Verlag, New York, pp. 307–309.

Volpon, A. G. T., De-Polli, H., and Döbereiner, J. 1981. Physiology of nitrogen fixation in *Azospirillum lipoferum* Br17 (ATCC 29709). *Arch. Microbiol.* 128:371–375.

Von Bülow, J. F. W., and Döbereiner, J. 1975. Potential for nitrogen fixation in maize genotypes in Brazil. *Proc. Nat. Acad. Sci. USA* 72:2389–2393.

Waksman, S. A. 1952. *Soil Microbiology*. John Wiley, New York, 347 p.

Walker, C. C., and Yates, M. G. 1978. The hydrogen cycle in nitrogen fixing *Azotobacter chroococcum. Biochimie* 60:225–232.

Wang, P. L., Koh, S. K., Chung, K. S., Uozumi, T., and Beppu, T. 1985. Cloning and expression in *E. coli* of the whole *nif* genes of *Klebsiella oxytoca*, a nitrogen fixer in the rhizosphere of rice. *Agric. Biol. Chem.* 49:1469–1477.

Wani, S. P., Dart, P. J., and Upadhyaya, M. N. 1983. Factors affecting nitrogenase activity (C_2H_2 reduction) associated with sorghum and millet established using the soil core assay. *Can. J. Microbiol.* 29:1063–1069.

Wani, S. P., Chandrapalaiah, S., and Dart, P. J. 1985. Response of pearl millet cultivars to inoculation with nitrogen-fixing bacteria. *Expl. Agric.* 21:175–182.

Watanabe, I., and Barraquio, W. L. 1979. Low levels of fixed nitrogen required for isolation of free-living N_2-fixing organisms from rice roots. *Nature* 277:565–566.

Watanabe, I., and Cabrera, D. R. 1979. Nitrogen fixation associated with the rice plant grown in water culture. *Appl. Environ. Microbiol.* 37:373–378.

Watanabe, I., and Lee, K. K. 1977. Non-symbiotic nitrogen fixation in rice and rice fields. *In* Ayanaba, A., and Dart, P. J. (eds.), *Biological Nitrogen Fixation in Farming Systems of the Tropics*. John Wiley and Sons, New York, pp. 289–305.

Watanabe, I., Lee, K. K., and Guzman, M. 1978. Seasonal change of N_2 fixing rate in rice field assayed by *in situ* acetylene reduction technique. II. Estimate of nitrogen fixation associated with rice plants. *Soil Sci. Plant Nutr.* 24:465–471.

Watanabe, I., Barraquio, W. L., Guzman, M. R., and Cabrera, D. A. 1979. Nitrogen-fixing (acetylene reduction) activity and population of aerobic heterotrophic nitrogen-fixing bacteria associated with wetland rice. *Appl. Environ. Microbiol.* 37:813–815.

Watanabe, I., Barraquio, W. L., and Daroy, M. L. 1982a. Predominance of hydrogen-utilizing bacteria among N_2-fixing bacteria in wetland rice roots. *Can. J. Microbiol.* 28:1051–1054.

Watanabe, I., Ventura, W., Cholitkul, W., Roger, P. A., and Kulasooriya, S. A. 1982b. Potential of biological nitrogen fixation in deepwater rice. *In* Proceedings of the International Deepwater Rice Workshop, Manila, 1981. Proceedings . . . Manila, IRRI, pp. 191–200.

Weier, K. L., MaCrae, I. C., and Whittle, J. 1981. Seasonal variation in the nitrogenase activity of a *Panicum maximum* var. Trichoglume pasture and identification of associated bacteria. *Plant Soil* 63:189–198.

Weller, D. M. 1984. Distribution of a take-all suppressive strain of *Pseudomonas fluorescens* on seminal roots of winter wheat. *Appl. Environ. Microbiol.* 48:897–899.

Wenzel, W., Singh, M., and Klingmüller, W. 1983. Molecular cloning of nitrogen fixation genes from *Azospirillum. In* Klingmüller, W. (ed.), Azospirillum *II: Genetics, Physiology, Ecology*. Birkhäuser, Basel, pp. 39–46 (Experientia supplementum, 48).

Westby, C. A., Cutshall, D. S., and Vigil, G. V. 1983. Metabolism of various carbon sources by *Azospirillum brasilense. J. Bacteriol.* 156:1369–1372.

Winogradsky, 1949. *Microbiologie du Sol.* Masson et Cie, Paris, 861 p.

Witty, J. F. 1979. Acetylene reduction assay can overestimate nitrogen-fixation in soil. *Soil Biol. Biochem.* 11:209–210.

Witty, J. F. 1983. Estimating N_2-fixation in the field using ^{15}N-labelled fertilizer: some problems and solutions. *Soil Biol. Biochem.* 15:631–639.

Witty, J. F., and Day, J. M. 1978. Use of $^{15}N_2$ in evaluating asymbiotic N_2 fixation. *In: Isotopes in Biological Dinitrogen Fixation.* International Atomic Energy Agency, Vienna, pp. 135–150.

Witz, D. F., Detroy, R. W., and Wilson, P. W. 1967. Nitrogen fixation by growing cells and cell-free extracts of *Bacillaceae. Arch. Microbiol.* 55:369–381.

Wong, P. P., and Stenberg, N. E. 1979. Characterization of *Azospirillum* isolated from nitrogen-fixing roots of harvested sorghum plants. *Appl. Environ. Microbiol.* 38:1189–1191.

Wong, P. P., Stenberg, N. E., and Edgar, L. 1980. Characterization of a bacterium of the genus *Azospirillum* from cellulolytic nitrogen-fixing mixed cultures. *Can. J. Microbiol.* 26:291–296.

Wood, A. G., Menezes, E. M., Dykstra, C., and Duggan, D. E. 1982. Methods to demonstrate the megaplasmids (or minochromosomes) in *Azospirillum. In* Klingmüller, W. (ed.), Azospirillum: *Genetics, Physiology, Ecology*. Birkhäuser, Basel, pp. 18–34 (Experientia supplementum, 42).

Wright, S. F., and Weaver, R. W. 1981. Enumeration and identification of nitrogen-fixing bacteria from forage grass roots. *Appl. Environ. Microbiol.* 42:97–101.

Wullstein, L. H., Bruening, M. L., and Bollen, W. B. 1979. Nitrogen fixation associated with sand grain root sheaths (rhizosheaths) of certain xeric grasses. *Physiol. Plant* 46:1–4.

Xavier, D. F., Baldani, J. I., Baldani, V. L. D., and Döbereiner, J. 1982. Effect of maize root extracts on growth and nitrogenase activity of *Azospirillum* spp. *An Acad. Bras. Cien.* 54:761.

Yao, P. Y., and Vincent, J. M. 1969. Host specificity in the root hair "curling factor" of *Rhizobium* spp. *Aust. J. Biol. Sci.* 22:413–423.

Yates, M. G., and Eady, R. R. 1979. The physiology and regulation of nitrogen fixation. *In* Subba Rao, N. S. (ed.), *Recent Advances in Biological Nitrogen Fixation.* Oxford and IBH Publishing, New Delhi, pp. 88–120.

Yates, M. G., Walker, C. C., Partridge, C. D. P., Pedrosa, F. O., Stephan, M. P., and Döbereiner, J. 1981. H_2 metabolism and nitrogenase activity in *Azotobacter chroococcum* and *Azospirillum brasilense*. *In* Gibson, A. G., and Newton, W. E. (eds.), *Current Perspectives in Nitrogen Fixation*. Academy of Sciences, Canberra, pp. 97–100.

Yoshida, T., and Ancajas, R. R. 1970. Application of the acetylene reduction method in nitrogen fixation studies. *Soil Sci. Plant Nutr.* 16:234–237.

Zimmer, W., Stephan, M. P., and Bothe, H. 1984. Denitrification by *Azospirillum brasilense* Sp7. I. Growth with nitrate as respiratory electron acceptor. *Arch. Microbiol.* 138:206–211.

Zöphel, G., Steinbauer, J., and Hess, D. 1980. *Azospirillum*: plasmid content and location of *nif* genes. First International Symposium on Bacteria–Plant Interactions, Bielefeld.

Zumft, W. G. 1985. Regulation of nitrogenase activity in anoxygenic phototrophic bacteria. *In* Evans, H. J., Bottomley, P. J., and Newton, W. E. (eds.), *Nitrogen Fixation Research Progress*. Martinus Nijhoff, Dordrecht, pp. 551–557.

Index

Page numbers in italics indicate primary references